Medani P. Bhandari

Roots to Revolution

The Transformation of Environmentalism

nova science publishers

www.novapublishers.com

DOI: https://doi.org/10.52305/BTTU0583

NOTICE TO THE READER

Library of Congress Cataloging-in-Publication Data

ISBN: 979-8-89530-382-5 (Hardcover)
ISBN: 979-8-89530-408-2 (eBook)

Published by Nova Science Publishers, Inc. † New York

To my friend, wife, coauthor Mrs. Prajita Bhandari, without your help and support this book was impossible

TOGETHER WE HAVE MILES TO GO

You have assumed a pivotal role akin to that of a true teacher, actively fostering an environmentally conscious atmosphere within your home, workplace, and community. Your impactful actions and exemplary behavior serve as an inspiration, motivating others to embrace sustainable practices and embrace a greener way of life. Your endeavors encompass a wide spectrum, from waste reduction to energy conservation and the advocacy for eco-friendly initiatives. Through these contributions, you actively participate in the creation of a better world for all. Your unwavering dedication to the stewardship of the environment shines as a guiding beacon, urging others to join the collective effort toward a healthier and more sustainable future. Amidst life's challenges, you provide me with a reassuring sense of comfort and a valuable lesson in resilience. For this, I extend my heartfelt gratitude to you.

Contents

Preface

Environmentalism has undergone a profound transformation over the last several decades, evolving from its early roots in conservation and preservation movements to an integrated, multifaceted approach addressing the global environmental crises of today. In this book, I aim to capture the depth and complexity of environmentalism as it has emerged and developed, drawing on over 30 years of research, activism, and scholarship. The goal is to provide a comprehensive understanding of where environmentalism has come from and offer insights into its future direction and the systemic changes needed to address the planet's unprecedented challenges. This study comprehensively explores *Roots to Revolution — The Transformation of Environmentalism*, addressing key objectives to guide global environmental efforts toward long-term sustainability.

Firstly, it seeks to define and articulate the concept of transformational environmentalism. The study emphasizes the need to move beyond incremental efforts and adopt systemic, large-scale transformations in today's global environmental crises. This shift focuses on addressing the root causes of environmental degradation and promoting long-term sustainability.

Next, the study explores the drivers of environmental degradation and the urgent need for systemic change. By identifying natural and human-induced factors—such as deforestation, fossil fuel consumption, and overexploitation—the research underscores the importance of addressing the structural drivers of environmental harm. It calls for comprehensive changes across sectors to move beyond short-term fixes.

A key contribution of this study is developing a comprehensive framework for systemic environmental change. By integrating policy, technology, social movements, and governance, this framework offers stakeholders a holistic and scalable approach to achieving transformational outcomes.

The study analyzes successful case studies of transformational environmental initiatives worldwide. These examples provide valuable insights into strategies that have worked and highlight lessons that can be applied or adapted to different contexts.

The research also delves into the social and political dimensions of transformational environmentalism. It explores how environmental issues intersect with social justice, equity, and political systems, advocating for inclusive approaches that address these disparities.

Finally, the study proposes strategies for overcoming barriers to change and offers a vision for the future of environmentalism, focused on achieving a sustainable, just, and resilient planet.

The scope of this study is global, encompassing the environmental, social, economic, and political dimensions of transformation. It examines how environmentalism has shifted in response to escalating ecological crises, such as climate change, biodiversity loss, and resource depletion. However, it further explores how these environmental issues intersect with social justice, cultural shifts, technological advancements, and global governance.

Historical Overview of Environmentalism — The opening chapter traces the historical development of environmentalism, from its early manifestations in the conservation movements of the 19th century to the modern-day climate action movement. It explores how environmentalism, rooted in a deep respect for nature, has morphed in response to industrialization, globalization, and increasing environmental degradation. By understanding the historical context, we gain insight into the evolving priorities and strategies that have shaped the movement.

Theoretical Foundations of Environmentalism — Environmentalism is grounded in a rich array of theoretical frameworks that inform our understanding of the relationship between humanity and the natural world. This chapter explores the philosophical underpinnings that have fueled the environmental movement, such as deep ecology, ecofeminism, and political ecology. It also delves into emerging systemic change and resilience theories, providing a foundational perspective for understanding how environmental transformation occurs.

Drivers of Environmental Transformation — Technological advancements, economic shifts, social movements, and political will drive environmental transformation. This section examines both the natural and human-induced drivers of environmental degradation, such as deforestation, fossil fuel dependency, and population growth, and the innovative drivers of change, including renewable energy, green technologies, and sustainable agriculture. Understanding these drivers is essential to framing the broader need for systemic transformation.

Key Stakeholders in Environmentalism Transformation — Environmentalism is not the domain of a single actor or group; it is a shared global responsibility. This chapter identifies the key stakeholders involved in environmental transformation, including governments, NGOs, businesses, local communities, and international organizations. It examines the roles each plays, their competing interests, and the collaborative efforts required to implement lasting change.

Pathways for Environmental Transformation — There are multiple pathways to achieving environmental transformation. This chapter explores strategies such as policy reforms, technological innovation, grassroots activism, and multilateral cooperation. Each path offers distinct advantages and challenges, but they form a roadmap for transitioning toward a sustainable and resilient planet.

Environmental Knowledge Systems and Innovation — The role of knowledge in shaping environmental transformation cannot be overstated. Here, we delve into the scientific, indigenous, and traditional knowledge systems that contribute to our understanding of environmental issues and solutions. Innovation is also explored, particularly how emerging technologies like AI, big data, and biotechnology shape environmental monitoring and management.

Environmentalism and Social Transformation — Environmentalism is inextricably linked with social transformation. The book highlights the intersectionality of environmental issues with social justice, equity, and human rights. This chapter addresses how environmental degradation disproportionately affects marginalized communities and how environmental justice movements seek to rectify these inequalities.

Challenges to Environmental Transformation — While widespread recognition of the need for environmental transformation exists, significant barriers remain. These include political resistance, economic interests, behavioral inertia, and social and psychological barriers to change. This chapter critically examines these challenges and offers strategies for overcoming them, including the role of education, policy change, and public engagement.

Environmentalism in a Globalized World — Environmental issues transcend national borders in today's globalized world. This chapter explores the implications of globalization on environmentalism, highlighting the interconnectedness of global trade, consumption patterns, and environmental impacts. It also examines international agreements, such as the Paris

Agreement and the Basel Convention, and the role of global governance in steering environmental action.

Case Studies of Environmental Transformation — Case studies offer practical insights into how environmental transformation can be achieved. This chapter presents examples worldwide, ranging from Costa Rica's conservation efforts to China's air pollution control strategies. These cases illustrate the successes and limitations of various approaches to environmentalism, providing valuable lessons for future endeavors.

Future of Environmentalism: Towards a Sustainable Planet — The book envisions a future where environmentalism evolves to meet the pressing challenges of the 21st century. It outlines the critical role of education, innovation, policy reforms, and social movements in accelerating environmental transformation. It also presents a vision for a sustainable future that is equitable, resilient, and just, emphasizing the need for holistic approaches that integrate environmental, social, and economic goals.

The final chapter synthesizes the key insights from the book and reaffirms the urgency of transformational environmentalism. It underscores the interconnected nature of environmental and social systems, arguing that true environmental transformation requires technical solutions and deep shifts in societal values, governance structures, and global cooperation.

I hope this book addresses the knowledge gap in transformational environmentalism, providing new insights and theories while paving the way forward to tackle the planetary crisis. Drawing from several decades of experience in environmental management, research, and study, I aim to illuminate the challenges we face and identify strategies for mitigation. I extend my heartfelt gratitude to my wife, Prajita Bhandari, for her unwavering support throughout our journey. Additionally, I would like to thank everyone who has encouraged me to pursue knowledge production.

Thank you!

Professor Medani P. Bhandari, PhD

Acknowledgments

The journey of writing *Roots to Revolution - The Transformation of Environmentalism* has been a deeply fulfilling and profoundly personal endeavor, reflecting a lifetime of academic exploration and intellectual growth. As I bring this project to fruition, I am grateful for the many individuals and experiences shaping this work.

First and foremost, I wish to express my heartfelt gratitude to my family, whose unwavering support and encouragement have been the foundation of my academic career. To my spouse, whose patience and understanding have been a constant source of strength, and to my children, whose curiosity and enthusiasm have inspired me every day—your love and support have made this journey possible.

I sincerely thank my mentors and colleagues who have guided and challenged me throughout my career. Your insights, feedback, and collaborative spirit have greatly enriched my understanding and have been instrumental in developing this book. Special thanks to Professor Steven R. Brechin, whose wisdom and encouragement have been invaluable, and to Syracuse University, NY, Brandeis University, MA, USA, ITC, the Netherlands, Tribhuvan University, Nepal, for the stimulating discussions and thoughtful critiques that have sharpened my perspective.

I am also grateful to the many scholars and researchers whose work has influenced and informed my own. The breadth and depth of knowledge in this field are vast, and the contributions of others have provided a rich tapestry of ideas upon which this book is built. To those whose writings and theories have inspired me, thank you for your intellectual generosity.

I extend my heartfelt gratitude to my dear friends, Mr. Kshitiz Raj Prasai, Rajan Adhikari, Dr. Medini Adhikari, Govinda Luitel, Binod Luitel and Professor Bishwa Kalyan Parajuli. Your steadfast encouragement and insistence on continuing my research and writing have been a resolute driving force, bolstering my conviction in this meaningful endeavor.

I sincerely thank Professor Mary Jo Bulbrook, PhD; Professor Ganesh Man Gurung, PhD; Professor Douglas Capogrossi, PhD; Professor Jacek Piotr Binda, PhD; Dr. Ambika Adhikari, PhD; Professor Keshav Bhattarai, PhD; Dr. Shvindina Hanna, PhD; Dr. Aleksander Łukasz Sapiński; Professor Durga Paudel, PhD; Mr. Jay Bhandari, and others for their encouragement and support in my academic pursuits.

"Janani Janmabhoomischa Swargadapi Gariyasi," a timeless adage that resounds with wisdom, acknowledges the unparalleled significance of both mother and motherland. At the outset, I extend my heartfelt gratitude to my mother, Heema Devi, whose uniqueness is a testament to the essence of all mothers, present and future, and Mother Earth herself. The depth of a mother's devotion is immeasurable, transcending enumeration. A mother's love remains matchless—until we fathom the depth of a mother's boundless, selfless affection, comprehending the very nature of maternal care, it is impossible to fully grasp the profound love and nurturing that Mother Earth extends to us and all living beings. Just as a mother's love is irreplaceable, so too is the immeasurable contribution of Mother Nature—a contribution that sustains and supports life in all its forms.

I would also like to extend my sincere gratitude to the invaluable contributions of my beloved family members, who have been unwavering pillars of support throughout my conservation journey. My wife, Prajita, along with my son, Prameya, daughter, Manaslu, and daughter-in-law, Kelsey Bhandari, son-in-law Abhimanyu, and his parents, Uma and Mahesh Iyer, thank you for your insightful comments and unwavering support during challenging times. To our grandchildren, Neena and Neelok, who brought joy to my family, I am grateful for your presence. I would also like to remember my two other grandchildren, Prasmin and Aahana Bhandari, and I hope that one day you will be reading this and other books of mine with pride.

My brothers Krishna and Shova, along with my nephews Jagannatha, Asmita, and Muna, brothers Hari and Durga, and daughter Pratigya, have been steadfast companions on this journey. My sisters Kali, Bhakti, Radha, Bindu, and Sita and their families have also provided unwavering support. Additionally, my uncle Bhim Prasad Bhandari and the late Goma Devi Bhandari, my siblings Govinda, Bishnu, Rudra, Bishma, and Sabita, together with their spouses and children, have been a constant source of strength and encouragement.

Prajita's parents, Dwarika Natha and Durga Devi, and her siblings Basudev, Ramesh, and Rishi, as well as Dhurba Acharya, Binod Ghimire, Ranjan Neupane, and Annapurna, Bijaya, Jaya, Ajaya, and their entire families have played an integral role in our endeavors. Without their unwavering support, my conservation journey would have faced interruptions or not have been possible. This sentiment applies universally—human beings thrive as social entities through mutual support.

A heartfelt and special acknowledgment is dedicated to my wife, Mrs. Prajita Bhandari, whose presence has created a tranquil space that enabled the completion of this book. Your perceptive insights and meticulous language editing have played an indispensable role in molding the outcome. You have woven new threads into our journey—our endeavors in conservation and humanitarianism—and your support has been my strength throughout the trials of my life. Words are inadequate to convey my gratitude, and it is worth noting that this book is not just my narrative but also a testament to the journey we are embarking upon together.

A special thanks to Nova Science Publishers, Inc. for your support and assistance in the publication process of this book.

And, to the future generations, whose well-being and future I hold close to my heart, thank you for motivating me to act and work towards a sustainable and thriving world for you.

Last but not least, I want to express my gratitude to all the readers. Your interest and engagement with my work fuels my passion for environmental conservation and social empowerment.

With a heart full of appreciation,

Thank you!

Professor Medani P. Bhandari, PhD

Abstract

Environmentalism has evolved from its nascent beginnings into a complex and multifaceted movement driven by an urgent need to address our planet's profound environmental crises. This book delves into the transformation of environmentalism, exploring its historical evolution, defining the essence of environmental transformation, and emphasizing the critical need for urgent change. By examining these foundational elements, we aim to provide a comprehensive understanding of how environmentalism can adapt and accelerate to meet the challenges of our time, ensuring our readers feel informed and knowledgeable.

The book's first section offers a historical overview of environmentalism, tracing its origins from early conservation efforts to the rise of modern environmental movements. This historical perspective provides context for understanding the evolution of environmental thought and action, highlighting key milestones and shifts in focus over time. We then define environmental transformation, examining what it entails and why it is essential for addressing the current and future environmental crises. The urgency of transformational change is underscored, reflecting the immediate need for comprehensive strategies to mitigate climate change, resource depletion, and ecosystem degradation.

Our study aims to outline the objectives and scope of environmental transformation, providing a roadmap for understanding the various dimensions of this complex issue. The book covers a wide range of topics, including theoretical foundations, drivers of transformation, key stakeholders, and pathways for change. We also explore environmental knowledge systems and innovation, the intersection of environmentalism and social transformation, and the challenges this endeavor faces.

Through case studies, we illustrate practical examples of environmental transformation worldwide, offering insights into successful strategies and innovative approaches. Finally, we look toward the future of environmentalism, discussing how emerging theories and practices can contribute to a sustainable planet. This book offers new perspectives and theories in environmentalism, aiming to accelerate the transformation needed to address the planetary crisis effectively.

Keywords: environmental transformation, environmental ethics, holistic environmentalism, systems thinking, global environmental crises, green transitions, environmental justice, environmental governance, UNFCCC, IPBES, indigenous communities, local actors, intermediaries, multipliers, environmental innovation, sustainable development goals, green economy, circular economy, renewable energy, biodiversity conservation, ecosystem services, decarbonization, carbon neutrality pathways, environmental knowledge, innovation, indigenous, local knowledge systems, democratization, big data, environmental monitoring, social transformation, grassroots activism, community empowerment, economic barriers, resource scarcity, access to resources, globalized world, europe, asia, latin america, north america, africa, middle east, planetary health

Chapter 1

Transformational Environmentalism: An Introuction

The concept of environmentalism has undergone significant transformation since its inception, evolving from a primarily nature-centered conservation movement to a multifaceted global initiative addressing complex ecological, social, economic, and political challenges. Early environmentalism was primarily driven by a desire to preserve pristine landscapes, protect biodiversity, and conserve natural resources, often in response to industrialization and urbanization's destructive impacts (Rootes, 2003; Otto, 2012; Eisner, 2007; Bothello et al., 2018). However, as the 20th century progressed, the environmental movement expanded its scope, embracing more comprehensive frameworks that acknowledged the interconnectedness of human society and natural ecosystems.

In the contemporary context, environmentalism has become a broad, inclusive philosophy that integrates sustainability, social justice, climate action, and environmental equity. This transformation reflects a shift from viewing environmental issues as isolated or regional concerns to recognizing them as global and systemic challenges requiring holistic, interdisciplinary approaches. The rise of climate change as a critical issue, along with the degradation of ecosystems, loss of biodiversity, and increased environmental inequities, has prompted the movement to adopt more inclusive and intersectional perspectives.

Technological advancements, globalization, and the democratization of information have also played significant roles in this transformation, enabling greater awareness, engagement, and activism. Environmentalism now intersects with human rights, gender equity, indigenous knowledge systems, and economic development, reflecting a growing recognition that ecological sustainability cannot be achieved without addressing the social and political dimensions of environmental issues.

Furthermore, the urgency of the environmental crisis has led to innovative solutions, from policy reforms like the Paris Agreement to grassroots movements advocating for climate justice. Environmentalism has evolved into a dynamic, adaptive movement that calls for systemic change, aiming to create a future where humans and the planet can thrive sustainably.

This transformation is not without its challenges. The movement faces the complex task of uniting diverse stakeholders, reconciling competing interests, and navigating a rapidly changing world's political, economic, and technological realities. Despite these obstacles, the evolution of environmentalism reflects a deeper understanding of the intricate relationship between humanity and the natural world, highlighting the need for collective action to address the pressing environmental issues of our time.

1.1. A Historical Overview of Environmentalism

Environmentalism, as a movement and ideology, has roots deeply embedded in humanity's evolving relationship with nature. Throughout history, people have exhibited concern for the environment, but the nature of these concerns has transformed dramatically based on societal needs, cultural values, and knowledge development. This overview traces the historical trajectory of environmentalism, exploring key stages, movements, and figures that have shaped its development (Barca, 2012; NA, 2016; Pádua, 2012; Carmichael et al., 2012) (Table 1).

Table 1. A historical timeline of environmentalism, highlighting key transformations and milestones

Time Period	Key Events/Transformations
Prehistory	Early human societies practiced sustainable living through hunting, gathering, and small-scale farming.
Ancient Civilizations (circa 3000 BCE – 500 CE)	Early awareness of nature's importance in civilizations like Mesopotamia, Egypt, and Greece influenced agricultural practices and city planning. The idea of "stewardship" began to appear in various cultural texts.
Middle Ages (500 – 1500)	Use of forests for fuel, timber, and agriculture intensified, especially in Europe. Concerns about deforestation and over-exploitation of natural resources arose, leading to early conservation efforts, such as forest laws in England.
Renaissance (1300 – 1600)	Technological innovations increased human control over nature. Exploration led to resource extraction from new lands and heightened awareness of natural wonders.
Industrial Revolution (1750 – 1850)	Major transformation in environmental impact due to the rise of factories, urbanization, and fossil fuel consumption. Air and water pollution became widespread, especially in Europe and North America. This period laid the foundation for the modern environmental movement.

Time Period	Key Events/Transformations
19th Century – Early Conservation Movement	Philosophers like Henry David Thoreau and naturalists like John Muir advocated for nature preservation. The U.S. created the National Park system (Yellowstone in 1872). Concerns about deforestation, wildlife conservation, and nature preservation grew.
Early 20th Century – Progressive Era (1900 – 1930)	Increased awareness of human environmental impact, mainly deforestation, water scarcity, and species extinction. Theodore Roosevelt's presidency saw major conservation efforts and the establishment of numerous national parks.
1930s – 1940s – Dust Bowl and WWII	The Dust Bowl in the U.S. led to soil conservation efforts. World War II increased industrial production, contributing to pollution and resource depletion. The end of the war sparked a renewed focus on environmental conservation.
1950s – Post-War Development	Rapid industrialization and urbanization led to growing environmental degradation. Concerns about nuclear fallout from weapons testing heightened awareness of human-induced ecological risks.
1960s – Rise of the Modern Environmental Movement	The publication of Rachel Carson's *Silent Spring* (1962) highlighted the dangers of pesticide use, sparking widespread public concern. The first Earth Day (1970) symbolized the birth of modern environmentalism. Environmental laws like the Clean Air Act and Clean Water Act were passed in the U.S.
1970s – Environmental Legislation and Global Awareness	Governments in the U.S. and Europe introduced major environmental legislation. International organizations like the United Nations launched initiatives focused on environmental sustainability. The Club of Rome's *Limits to Growth* report (1972) raised global awareness about sustainability.
1980s – Globalization and Environmental Crises	Major environmental disasters, like the Bhopal disaster (1984) and the Chornobyl nuclear accident (1986), drew global attention. The Brundtland Commission's report *Our Common Future* (1987) introduced the concept of sustainable development.
1990s – Climate Change and Biodiversity Concerns	The 1992 Earth Summit in Rio created key treaties on climate change, biodiversity, and desertification. The Kyoto Protocol (1997) set binding emission reduction targets for industrialized countries, marking a significant moment in climate policy.
2000s – Climate Action and Environmental Justice	Growing concerns about global warming led to worldwide movements to combat climate change. Environmental justice became a major theme, addressing the unequal impacts of environmental degradation on marginalized communities. Movements like the global anti-globalization protests brought attention to the environmental costs of free trade.
2010s – Climate Agreements and Technological Innovations	The Paris Agreement (2015) committed nations to limit global warming to below 2°C. Technological advances in renewable energy, electric vehicles, and sustainable agriculture gained momentum. The rise of youth movements, like Fridays for Future, emphasized the urgency of climate action.
The 2020s – Planetary Crisis and Transformation	The COVID-19 pandemic highlighted the links between environmental degradation and human health. The concept of the planetary crisis emerged, urging urgent action on climate change, biodiversity loss, and sustainability. Global movements like Extinction Rebellion pushed for systemic change. Efforts to integrate traditional and indigenous knowledge in environmental solutions gained prominence.

Source: Author compilation.

This timeline provides an overview of the evolution of environmentalism, showcasing key moments of transformation that shaped the movement over time.

1.1.1. Pre-Industrial Roots: Nature in Myth, Religion, and Early Conservation (Ancient Times – 18th Century)

Before the emergence of formal environmental movements, respect for nature was often embedded in myth, religion, and traditional practices. Many ancient civilizations and indigenous peoples recognized the importance of living harmoniously with the natural world (Coates, 2013; Stephan and Stephan, 2015; Roux et al., 2022). Examples include:

- *Indigenous Knowledge Systems*: Indigenous communities across the Americas, Africa, and Asia practiced sustainable land use, fishing, and agriculture, emphasizing the stewardship of nature for future generations. Their deep ecological knowledge was built on reciprocity with the environment.
- *Religious Traditions*: Hinduism, Buddhism, and Taoism, among other faiths, incorporated elements of environmental care into their doctrines, stressing the sacredness of nature. Ancient texts like the *Rigveda* and *Tao Te Ching* reflect a reverence for nature and advocate for ethical living in the environment.

However, with the rise of agricultural societies and later urbanization, environmental degradation became more common. As populations grew, deforestation, overgrazing, and soil depletion became widespread. The connection between human well-being and environmental health was understood in localized ways, with some early responses focused on resource management, such as establishing hunting preserves or forest reserves by monarchies in Europe and Asia.

1.1.2. The Industrial Revolution and the Birth of Modern Environmentalism (Late 18th Century – Early 20th Century)

The Industrial Revolution (beginning in the late 18th century) marked a turning point in humanity's relationship with the environment. Industrialization led to unprecedented economic growth, urbanization, and technological progress, but it also caused significant environmental degradation, including air and water pollution, deforestation, and the destruction of natural habitats (Granjou, 2016; Stobart, 2013; Zhu and Wang, 2020; McBeath and Rosenberg, 2006; Nanda and Pring, 2012; Reno, 2020).

- *Romanticism and Early Environmental Thought*: The Romantic movement of the late 18th and early 19th centuries emerged partly as a reaction to industrialization. Thinkers like William Wordsworth and Ralph Waldo Emerson emphasized nature's spiritual and aesthetic value, offering a counter-narrative to the utilitarian view of nature as a resource for human exploitation.
- *Conservation and Preservation Movements*: By the late 19th century, environmental concerns began to manifest more formally in the form of the conservation and preservation movements. The conservation movement, led by figures such as Gifford Pinchot, advocated for the sustainable use of natural resources to ensure their availability for future generations. In contrast, the preservation movement, championed by John Muir, sought to protect wilderness areas from human exploitation altogether. Muir's efforts led to creating national parks in the United States, with Yellowstone becoming the first national park in 1872.
- *Transcendentalism:* The 19th-century transcendentalist movement, which included figures like Henry David Thoreau, emphasized a spiritual connection to nature and promoted the idea of simple living in harmony with the environment. Thoreau's *Walden* (1854) is considered a foundational text of American environmental thought.

1.1.3. The Progressive Era and Early Environmental Reforms (Early 20th Century)

The early 20th century saw the rise of the Progressive Era, a period of social and political reform that addressed issues related to public health, urbanization, and environmental conservation. This era laid the groundwork for future environmental regulations and public policies (Granjou, 2016; Stobart, 2013; Zhu and Wang, 2020; McBeath and Rosenberg, 2006; Nanda and Pring, 2012; Reno, 2020).

- *Theodore Roosevelt's Conservation Efforts:* U.S. President Theodore Roosevelt was a major figure in the conservation movement, establishing numerous national parks, forests, and wildlife refuges. He worked closely with Gifford Pinchot to promote the idea that natural resources should be used responsibly for society's benefit (Brown, 2007).
- *The Dust Bowl:* During the 1930s, the American Midwest experienced severe dust storms and soil erosion, leading to widespread agricultural collapse. This environmental disaster highlighted the need for better land management practices and spurred federal efforts like establishing the Soil Conservation Service (Hansen and Libecap, 2004).

1.1.4. The Post-War Era and the Rise of Modern Environmentalism (1940s – 1960s)

In the aftermath of World War II, the rapid industrialization and economic growth that followed led to new environmental challenges, including pollution, waste, and habitat destruction. At the same time, public awareness of environmental issues began to grow, setting the stage for the environmental movement of the 1960s (Chung, 1997; Robertson et al., 2020; McBeath and Rosenberg, 2006).

- *Aldo Leopold and Ecocentric Thought:* Aldo Leopold's *A Sand County Almanac* (1949) is a key text that introduced the concept of the "land ethic. " It argues that humans should see themselves as part of the biotic community, responsible for caring for the environment. This ecocentric perspective helped shape modern environmental ethics (Kuhlmann, 1996).
- *Rachel Carson and the Environmental Movement:* The publication of Rachel Carson's *Silent Spring* in 1962 is often regarded as the beginning of the modern environmental movement. Carson's work exposed the dangers of pesticide use, particularly DDT, on ecosystems and human health, galvanizing public concern and leading to increased regulatory efforts. Her work also played a significant role in the eventual establishment of the U.S. Environmental Protection Agency (EPA) in 1970 (Lytle, 2007).

1.1.5. The Environmental Movement of the 1960s and 1970s

The environmental movement gained momentum in the 1960s and 1970s, driven by growing public awareness of pollution, environmental degradation, and the need for action to address these issues. This period saw the formation of key environmental organizations, new legislation, and global awareness of environmental challenges.

- *Earth Day and Legislative Achievements*: The first Earth Day was celebrated on April 22, 1970, marking a turning point in public engagement with environmental issues. This period also saw the passage of landmark environmental legislation in the United States, including the Clean Air Act (1970), the Clean Water Act (1972), and the Endangered Species Act (1973) (Johnson and Greenberg, 2017).
- *International Cooperation*: Environmentalism became a global issue in the 1970s, leading to the first United Nations Conference on the Human Environment in Stockholm in 1972. This conference marked the beginning of international environmental governance and highlighted the need for global cooperation to address environmental problems (Handl, 2012).

1.1.6. Environmentalism and Sustainable Development (1980s – Present)

By the 1980s, environmentalism had evolved to encompass broader concerns about sustainable development, climate change, and environmental justice. The concept of sustainable development, popularized by the Brundtland Report in 1987, emphasized the need to balance economic growth, environmental protection, and social equity (Adams, 2005; Redclift, 2005; Nygren, 1998).

- *Climate Change and Global Action:* Scientific research and international efforts, such as the Kyoto Protocol (1997) and the Paris Agreement (2015), increased attention to climate change in the 1990s and early 2000s. Climate change became a central focus of the environmental movement, with activists, scientists, and policymakers working to address the global impacts of greenhouse gas emissions and rising temperatures (Bhandari, 2022).

- *Environmental Justice and Intersectionality:* In recent decades, environmentalism has increasingly intersected with social justice issues. The environmental justice movement, which emerged in the 1980s, focuses on addressing marginalized communities' disproportionate environmental burdens, particularly concerning pollution, waste disposal, and access to clean water and air. The movement has expanded to include gender, racial, and economic justice as integral aspects of environmental advocacy (Di Chiro, 2020).

The history of environmentalism is one of evolving concerns and strategies shaped by changing societal values, scientific discoveries, and ecological realities. From its early roots in conservation and preservation, environmentalism has grown into a global movement that addresses various issues, from climate change and biodiversity loss to environmental justice and sustainable development. As the 21st century progresses, environmentalism transforms, adapting to new challenges and pushing for systemic changes to ensure the planet's and its inhabitants' long-term health and well-being.

1.2. Defining Environmental Transformation

Environmental transformation refers to profound and systemic changes in the relationship between human societies and the natural environment. It encompasses shifts in ecological conditions, societal values, technological advancements, governance structures, and the global economy. Environmental transformation is not limited to specific environmental issues like pollution or climate change; it is a broad, multi-dimensional concept that recognizes the interconnectedness of ecological, social, economic, and political systems (Feola, 2015; O'Brien, 2012; Hurlimann et al., 2021).

The term can be understood as a natural process—where ecosystems change due to natural forces such as geological, climatic, and biological factors—and a human-induced phenomenon, where human activity alters local, regional, and global environmental systems. Today's key drivers of environmental transformation include climate change, biodiversity loss, technological advancements, urbanization, industrialization, and societal awareness and governance shifts.

1.2.1. Key Components of Environmental Transformation

Environmental transformation is marked by several critical elements that reflect its comprehensive nature. These elements illustrate how various interconnected factors drive transformations (Feola, 2015; Elliot, 2011; Boström and Lidskog, 2024).

- *Ecological and Climate Shifts:* The most direct and observable component of environmental transformation is the change occurring within ecosystems and the planet's climate system. This aspect includes:
- *Climate Change:* The rapid increase in global temperatures due to human-induced greenhouse gas emissions is one of the most significant examples of environmental transformation. Climate change alters weather patterns, sea levels, and ecosystems, leading to long-term impacts on the environment and human societies. Extreme weather events, desertification, rising sea levels, and changes in biodiversity are all linked to this process (Boström and Lidskog, 2024).
- *Biodiversity Loss*: Human activity has significantly impacted global biodiversity through deforestation, habitat destruction, pollution, and overexploitation of resources. These pressures are transforming ecosystems, from rainforests to coral reefs, and causing species extinction at unprecedented rates.
- *Land Use Changes:* Agricultural expansion, urbanization, and deforestation have transformed landscapes globally. These changes have major environmental consequences, including soil degradation, water resource depletion, and altered ecosystems.

1.2.2. Technological Advancements and Innovations

Technological advancements play a dual role in environmental transformation. On the one hand, technology can be a driver of environmental degradation; on the other hand, technology innovation is crucial for solving environmental challenges and transforming how societies interact with the natural world (Mehmood et al., 2024; Zhang et al., 2022; Shao and Chen, 2022).

- *Industrialization and Environmental Impact*: The Industrial Revolution marked a pivotal shift in environmental transformation, as rapid industrial growth led to increased pollution, resource depletion, and habitat destruction. The reliance on fossil fuels and large-scale manufacturing has had long-term environmental consequences, such as air and water pollution and climate change.
- *Green Technology and Sustainability*: The 21st century has seen a surge in the development of technologies to mitigate environmental degradation. Renewable energy technologies (e.g., solar, wind, hydropower), sustainable agriculture practices, and green transportation systems transform how human societies function harmoniously with the environment. These technologies aim to decouple economic growth from environmental harm, promoting a transition toward sustainability.

1.2.3. Societal and Cultural Shifts

Environmental transformation is also influenced by changing societal values and cultural perspectives toward nature, sustainability, and equity. Social awareness of environmental issues has increased significantly in recent decades, and this shift is shaping policy, business practices, and individual behavior (Plater, 2005; Banerjee, 2001).

- *Environmental Awareness and Movements*: The rise of global environmental movements, such as those sparked by Rachel Carson's publication of Silent Spring and the first Earth Day in 1970, has played a key role in transforming public perceptions of environmental issues. More recently, movements like Fridays for Future and Extinction Rebellion have catalyzed a new generation of environmental activism, highlighting the urgency of climate action.
- *Consumer and Corporate Responsibility*: The transformation of societal values has also led to changes in consumer behavior and corporate practices. Ethical consumption, where consumers choose products based on their environmental and social impacts, has gained traction. Businesses are increasingly held accountable for their environmental practices, with concepts like corporate social responsibility (CSR) and environmental, social, and governance

(ESG) standards becoming mainstream (Karwowski and Raulinajtys-Grzybek, 2021).

1.2.4. Political and Governance Changes

Governance and political structures are key to driving environmental transformation. Governments, international organizations, and non-governmental entities shape policies that affect the environment, natural resources, and sustainability efforts (Bernstein, 2002; Haq and Paul, 2013; Arts and Leroy, 2006).

- *Environmental Governance*: Over the past few decades, international environmental governance has expanded through treaties, conventions, and global frameworks to address pressing environmental challenges. Examples include the Paris Agreement on climate change, the Convention on Biological Diversity, and the United Nations Framework Convention on Climate Change (UNFCCC) ((Boström and Lidskog, 2024).
- *Decentralized and Community-Based Management*: The shift toward decentralized environmental governance, where local communities play a more active role in managing their resources, has emerged as an important aspect of environmental transformation. In many parts of the world, indigenous and local communities are leading conservation and sustainable management efforts, using traditional knowledge alongside modern techniques.

1.2.5. Economic Shifts and Sustainable Development

The global economy is transforming industries, and governments recognize the need for sustainable development practices that do not compromise the planet's ability to sustain life.

- *Green Economy:* A green economy is defined as one that improves human well-being and social equity while significantly reducing environmental risks and ecological scarcities. By incorporating sustainability principles, transitioning to a green economy transforms

traditional industries such as energy, manufacturing, and agriculture (Adams, 2005).
- *Sustainable Development Goals (SDGs):* The adoption of the United Nations' Sustainable Development Goals (SDGs) in 2015 reflects a global effort to balance economic development with environmental protection and social equity. As outlined in the SDGs, environmental transformation includes reducing carbon emissions, protecting biodiversity, and promoting sustainable resource use (Davey, 2009).

1.2.6. Social and Environmental Justice

Environmental transformation increasingly involves addressing inequities in how environmental harms and benefits are distributed across different populations. Historically, marginalized communities—predominantly Indigenous peoples, low-income groups, and communities of color—have disproportionately borne the brunt of environmental degradation (Martínez-Alier, 2020; Rios, 2018; Gilmurray, 2018).

- *Environmental Justice:* The environmental justice movement seeks to rectify these imbalances by advocating for fair treatment and meaningful involvement of all people in environmental decision-making processes. This movement emphasizes that the benefits of environmental protection, such as clean air and water, should be shared equally and that the burdens of environmental degradation should not disproportionately fall on marginalized communities.
- *Intersectionality and the Environment:* Intersectional approaches to environmentalism explore how social categories like race, class, gender, and nationality intersect with environmental issues. Environmental transformation increasingly requires considering the complex social dynamics that underlie issues such as access to natural resources, exposure to pollution, and vulnerability to climate impacts.

1.2.7. Scientific Discoveries and Ecological Understanding

The advancement of scientific knowledge has transformed humanity's understanding of the environment, contributing to more informed decision-making and policy formulation (Pepper, 2019).

- *Earth Systems Science:* The development of earth systems science has provided a more holistic understanding of how various components of the earth—the atmosphere, biosphere, hydrosphere, and geosphere—interact. This science has underscored the global nature of environmental problems, such as climate change and biodiversity loss, and the importance of integrated solutions (Patten, 2014).
- *Sustainability Science*: Sustainability science has emerged as an interdisciplinary field focused on the interaction between natural and social systems. Its aim is to advance knowledge that can lead to solutions for sustainable development. It bridges ecology, economics, sociology, and political science, contributing to comprehensive strategies for environmental transformation (Bhandari, 2019).

Environmental transformation reflects the ongoing and dynamic shifts in how human societies interact with and impact the natural world. It involves ecological, technological, societal, and political changes, encompassing a wide range of issues from climate change and biodiversity loss to social justice and sustainable development. As humanity grapples with the environmental challenges of the 21st century, the transformation of environmental systems and the corresponding human responses will determine the future trajectory of both the planet and its inhabitants (Pepper, 2019; Patten, 2014; Passmore, 2017).

1.3. The Urgency of Transformational Change in Environmentalism

Environmentalism, as a movement, has long called for changes in human behavior and policies to protect the natural world. However, given the accelerating pace of environmental degradation and the mounting evidence of irreversible harm, the call for transformational change has become more urgent than ever. Transformational change in environmentalism refers to profound, systemic shifts beyond incremental adjustments, aiming to overhaul the structures, behaviors, and systems that drive environmental harm. The need for such comprehensive change is underscored by the severity and interconnectedness of today's environmental crises, including climate change, biodiversity loss, resource depletion, pollution, and environmental injustice (O'Brien, 2012; Feola, 2015; Grin et al., 2010).

1.3.1. Global Environmental Crises: The Scale of the Problem

The scale and intensity of global environmental crises demand urgent action. While valuable, incremental efforts are no longer sufficient to address the planet's breadth of challenges (Smil, 1987; Lomborg, 2004; Owen et al., 2006). The following crises highlight the need for immediate and deep systemic transformation:

- *Climate Change:* The Earth's climate is changing at an unprecedented rate due to human activities, primarily the burning of fossil fuels and deforestation. The 2021 report by the Intergovernmental Panel on Climate Change (IPCC) starkly warned that without rapid and large-scale changes, the global temperature rise will exceed 1.5°C above pre-industrial levels, leading to catastrophic consequences. These include frequent and intense heatwaves, storms, droughts, rising sea levels, and ecosystem collapse. Small-scale reforms in emissions reductions are inadequate to curb the trajectory; transformative action is necessary to decarbonize economies entirely (Bhandari 2024).
- *Biodiversity Loss and Ecosystem Collapse*: According to the Intergovernmental Science-Policy Platform on Biodiversity and Ecosystem Services (IPBES), approximately 1 million species face extinction within decades, primarily due to human activities like habitat destruction, pollution, overexploitation, and climate change. The degradation of ecosystems threatens their services, from pollination and water purification to climate regulation and food security. The current approach to conservation—protecting isolated areas or species—is insufficient given the rapid pace of habitat destruction; a transformative strategy that integrates biodiversity protection into every aspect of human activity is crucial.
- *Resource Depletion and Overconsumption*: The global economy operates on an unsustainable resource model that extracts, consumes, and wastes resources at an alarming rate. The concept of Earth Overshoot Day—the date when humanity's demand for ecological resources exceeds what the Earth can regenerate in a year—illustrates how human consumption outstrips planetary boundaries. To prevent the collapse of natural systems that sustain human life, transformative changes in production, consumption, and waste management are essential, including shifts toward circular economies and sustainable resource use.

1.3.2. Limitations of Incremental Change

While incremental changes—such as adopting energy-efficient technologies, passing environmental regulations, or promoting conservation practices—are important, they are not sufficient to address the depth and urgency of the global environmental crises (Bothello and Salles-Djelic, 2018; Geels, 2011; Heucher et al., 2024). There are several reasons why these gradual approaches are falling short:

- *Systemic Problems:* Many environmental issues are symptoms of deeper systemic problems rooted in industrialization, capitalism, unsustainable economic growth models, and social inequities. For example, simply improving the efficiency of coal power plants must address the fact that fossil fuel dependency is the root cause of climate change. A transformational shift to renewable energy and sustainable development is necessary to tackle these systemic drivers.
- *Ecological Tipping Points*: Many ecosystems are approaching tipping points, beyond which they will no longer function in ways that support human life. For instance, the collapse of coral reefs, loss of Arctic ice, or disruption of the Amazon rainforest's water cycle could have global ramifications. Given the speed at which these tipping points are approaching, incremental policy reforms that take years or decades to implement will likely arrive too late to prevent irreparable damage.
- *Pace of Change vs. Pace of Crisis*: Environmental degradation is accelerating faster than the implemented solutions. For instance, while renewable energy adoption is increasing, fossil fuel consumption continues to grow in many regions. The gradual phasing out of harmful technologies and practices must be replaced by bold, large-scale shifts that can keep pace with the escalating environmental crises.

1.3.3. What Transformational Change Entails

Transformational change in environmentalism is about creating shifts across multiple dimensions—ecological, social, economic, political, and cultural—by rethinking how humanity interacts with the natural world (Ayars, 2009;

Henderson, 2002; Deubelli and Mechler, 2021; Landau, 2005). The following elements highlight the critical aspects of such transformative shifts:

- *Deep Decarbonization:* A core component of transformational environmentalism is shifting from fossil fuels to 100% renewable energy. This transition involves changing the energy sector and rethinking economic systems dependent on carbon-based fuels. Policies must accelerate the deployment of renewable energy technologies, electrify transportation and industry, promote energy efficiency, and ensure a just transition for workers in carbon-intensive sectors (Geels et al., 2017).
- *Circular Economy:* Moving from a linear economy (based on a take-make-dispose model) to a circular economy is crucial for reducing resource depletion and minimizing waste. A circular economy focuses on designing out waste, keeping products and materials in use for as long as possible, and regenerating natural systems. This involves radical changes in manufacturing processes, product design, recycling systems, and consumer behavior, all aimed at reducing environmental impacts across the entire life cycle of products (Corvellec et al., 2022).
- *Ecosystem Restoration and Regenerative Practices:* Besides preventing further damage, transformational change in environmentalism emphasizes restorative and regenerative practices. These include large-scale reforestation, wetland restoration, and regenerative agriculture, which enhance soil health, sequester carbon, and promote biodiversity. Ecosystem restoration helps reverse the damage caused by decades of unsustainable land use and offers opportunities to enhance climate resilience (Morseletto, 2020).
- *Environmental Justice and Equity:* Transformational change must address the social dimensions of environmental issues, particularly the unequal distribution of environmental harms and benefits. Marginalized communities often bear the brunt of pollution, resource extraction, and climate impacts while contributing the least to these problems. A just transition ensures that climate policies and environmental solutions do not perpetuate inequalities (Larkins, 2024). It also involves amplifying the voices and leadership of indigenous peoples and marginalized communities in environmental governance.

- *Behavioral and Cultural Shifts:* Transformational change requires technological solutions and profound cultural and behavioral shifts. This involves rethinking how societies value nature, moving away from viewing the environment as an endless resource toward seeing it as a system that humans are part of and dependent on (Luo, 2024). Public education, cultural change, and behavioral interventions are key in reshaping individual and collective values and practices to align with sustainable living.

1.3.4. The Role of Policy and Governance in Driving Transformation

Governments, institutions, and multilateral organizations are critical in driving transformational environmental change. Policy interventions must go beyond short-term fixes and instead focus on systemic reform and long-term sustainability goals (Purcell and Chahine, 2019; Klijn, 2004; Patterson et al., 2017). Critical aspects of policy-driven transformation include:

- *International Agreements and Cooperation*: The global nature of environmental challenges requires international collaboration. Agreements such as the Paris Agreement and the Convention on Biological Diversity aim to coordinate global action, but more ambitious targets and compliance mechanisms are necessary to enforce meaningful change. International cooperation can also facilitate technology transfer, capacity building, and financial support for developing nations transitioning to sustainable practices.
- *Bold National and Regional Policies*: Countries and regions must implement ambitious policies that accelerate the transition to sustainability. Examples include the European Green Deal, which aims for Europe to become climate-neutral by 2050, and national climate plans that target specific sectors such as energy, agriculture, and transportation. Governments can use carbon pricing, emissions reduction mandates, and renewable energy subsidies to push for systemic change.
- *Decentralized Governance and Local Action:* Transformational environmental change also requires the engagement of local governments and communities. Decentralized governance models that empower cities, regions, and local populations to take action can be more adaptive and responsive to local environmental challenges.

Community-driven initiatives, especially those rooted in indigenous knowledge and practices, can offer innovative and context-specific solutions to environmental problems.

1.3.5. Barriers to Transformational Change

While the urgency for transformational environmentalism is clear, there are significant barriers that need to be overcome to implement it effectively (Navickaite and Janiunaite, 2012):

- *Political Resistance:* In many regions, vested interests, particularly from the fossil fuel industry and other carbon-intensive sectors, resist transformational change. Political lobbying, misinformation campaigns, and short-term economic interests often hinder policy progress on climate action and environmental protection.
- *Economic Dependence on Unsustainable Industries*: Many economies, especially in developing regions, rely heavily on industries such as mining, agriculture, and fossil fuel extraction. Transformational change must offer alternative economic pathways that provide jobs and livelihoods while promoting sustainability. The fear of economic dislocation or job losses is a major barrier to adopting aggressive environmental policies.
- *Societal Inertia and Cultural Resistance*: Cultural norms and consumption patterns in many parts of the world are deeply ingrained and difficult to shift. The expectation of limitless growth, resource consumption, and disposable lifestyles challenges transformational change. Public education, awareness, and cultural shifts are essential for fostering broader acceptance of sustainable practices.

The urgency of transformational change in environmentalism cannot be overstated. The current trajectory of environmental degradation threatens the stability of ecosystems, human health, and future generations. Incremental solutions, while necessary, are insufficient to address the complex, interconnected, and rapidly worsening environmental crises. To secure a sustainable future, bold, systemic, and transformative changes are required across all sectors of society, involving technological innovation, governance reform, economic restructuring, and cultural shifts(Duffy, 2009; Navickaite and Janiunaite, 2012; Kezar, 2013). Only through deep transformation can we hope to reverse the damage done to the planet and build a future where both people and nature can thrive.

1.4. Objectives and Scope of the Study

The book Transformational Environmentalism aims to delve into the critical environmental degradation issues and explore the pathways to profound systemic change. Given the increasing urgency of global environmental crises, the book seeks to offer a comprehensive framework for understanding how societies can transition from a state of environmental exploitation and unsustainable practices to a more harmonious relationship with the natural world. The study is rooted in the belief that incremental efforts are insufficient, and only through transformational change can humanity avert ecological collapse and secure a sustainable future.

1.4.1. Objectives of the Study

The primary objectives of the study are centered around understanding, analyzing, and proposing solutions for the transformational changes required in environmentalism:

1.4.1.1. To Define and Articulate the Concept of Transformational Environmentalism

One key objective is to provide a clear and robust definition of transformational environmentalism. This involves distinguishing between incremental reforms and transformational shifts fundamentally altering the structures and systems that drive environmental degradation. The book seeks to establish the theoretical foundations of transformational environmentalism, exploring how radical changes in governance, economics, culture, technology, and social behavior are necessary to create a sustainable future.

1.4.1.2. To Explore the Drivers of Environmental Degradation and the Need for Systemic Change

The study aims to examine the root causes of environmental degradation, looking beyond surface-level issues like pollution or deforestation to explore the deeper systemic drivers such as industrialization, capitalism, and unsustainable growth models. By identifying these underlying causes, the book will argue that piecemeal solutions are insufficient and that fundamental transformations are required in key areas such as energy, agriculture, transportation, consumption, and land use.

1.4.1.3. To Provide a Comprehensive Framework for Systemic Environmental Change

The book will offer a detailed framework that outlines the steps and actions necessary to achieve transformational environmental change. This includes a focus on the interconnectedness of ecological, economic, political, and social systems, and the need for a holistic approach that integrates these dimensions. It will propose pathways for transitioning to a green economy, adopting sustainable technologies, protecting biodiversity, and ensuring environmental justice.

1.4.1.4. To Analyze Successful Case Studies of Transformational Environmental Initiatives

The study aims to identify and analyze successful examples of transformational environmental change worldwide. These case studies will span different sectors, geographies, and scales, including national policies, community-driven initiatives, and corporate transformations. By studying these examples, the book seeks to provide inspiration, guidance, and practical lessons for others aiming to implement similar transformations.

1.4.1.5. To Address the Social and Political Dimensions of Transformational Environmentalism

Environmental change cannot be separated from social justice, equity, and governance issues. One objective of the book is to explore how transformational environmentalism must address issues of inequality, particularly how marginalized communities are disproportionately affected by environmental harm and excluded from decision-making processes. The study will also focus on political challenges, exploring how power dynamics, vested interests, and resistance to change act as barriers to transformational progress.

1.4.1.6. To Propose Strategies for Overcoming Barriers to Transformational Change

A key objective is identifying and proposing strategies for overcoming the barriers hindering transformational change. These barriers include political resistance, economic inertia, cultural attitudes, and technological limitations. By analyzing these obstacles, the book will offer practical solutions for overcoming them, including policy recommendations, technological innovations, educational reforms, and strategies for building public support and political will for environmental action.

1.4.1.7. To Provide a Vision for the Future of Environmentalism

Finally, the book aims to provide a visionary outlook on the future of environmentalism. It will explore how humanity can shift toward a regenerative and restorative relationship with nature, where sustainability is embedded into every aspect of life. This vision includes environmental sustainability, social well-being, equity, and intergenerational responsibility, offering a roadmap for achieving a thriving planet that can support future generations.

1.4.2. Scope of the Study

The scope of "Transformational Environmentalism" is broad, as it seeks to address multiple dimensions of the environmental crisis and the need for systemic change. The interdisciplinary study integrates insights from ecology, economics, political science, sociology, ethics, and technology. It explores theoretical concepts and practical solutions, making it relevant to scholars, policymakers, activists, and practitioners. The book covers the following key areas:

1.4.2.1. Global Environmental Crises and Challenges

The scope of this study encompasses several major global environmental crises currently threatening the planet. Climate change is a key focus, examining the impacts of global warming, rising sea levels, extreme weather events, and the critical need for decarbonization to mitigate these effects. Biodiversity loss is another pressing issue, highlighting the mass extinction of species and the collapse of ecosystems essential for life on Earth.

The study also addresses pollution and resource depletion, including air and water pollution, land degradation, and the unsustainable extraction of natural resources, which further exacerbate environmental decline. Additionally, it emphasizes environmental justice, recognizing that marginalized and vulnerable communities disproportionately suffer the consequences of environmental degradation.

By focusing on these interconnected crises, the study provides a comprehensive understanding of the urgent need for transformational change to safeguard the planet and promote a more just and sustainable future for all.

1.4.2.2. Theoretical Frameworks and Concepts of Transformational Change

The book delves into the theoretical foundations of transformational environmentalism, exploring key concepts critical to understanding and addressing global environmental challenges. One of the central themes is ecological limits and planetary boundaries. This concept outlines the thresholds within which humanity must operate to avoid causing irreversible environmental damage. The study underscores the importance of maintaining balance within Earth's natural systems by defining these limits.

Sustainable development and the circular economy are also key areas of focus. The book examines how societies and economies can transition toward sustainability by adopting circular economic models prioritizing resource efficiency, waste reduction, and material recycling. These models, alongside sustainable development practices, provide a roadmap for reducing environmental impact while supporting economic growth.

A systems-thinking approach is essential for understanding environmental issues' complexity. This perspective considers the interconnectedness of ecological, social, and economic factors, offering a holistic view of the challenges and solutions needed to address environmental degradation.

The book explores regenerative and restorative practices, which go beyond sustainability by actively healing and revitalizing ecosystems. It discusses initiatives like rewilding, ecosystem restoration, and sustainable agriculture, highlighting how these practices can restore damaged environments and create a net positive impact on the planet. Through these concepts, the book offers a comprehensive framework for transformational environmentalism to foster a sustainable and regenerative future.

1.4.2.3. Technological and Economic Transformations

The study's scope includes a comprehensive analysis of the technological and economic transformations necessary to tackle the ongoing environmental crises. One of the primary areas of focus is renewable energy and decarbonization. The study explores the transition from fossil fuels to renewable energy sources like wind, solar, and hydropower. It also addresses the challenges of achieving energy sustainability, including infrastructure, costs, and scaling these technologies to meet global energy demands.

Another key area is green technologies and innovation. The book examines the role of emerging technologies, including electric vehicles, smart grids, and carbon capture, in driving environmental progress. These innovations are critical to reducing emissions and enhancing energy efficiency, offering new ways to mitigate the effects of climate change and pollution.

The study also emphasizes the importance of the circular economy and resource efficiency. It explores strategies for minimizing waste, promoting recycling, and creating closed-loop systems that reduce resource extraction and environmental impact. By encouraging more efficient use of materials, these approaches support a more sustainable economic model.

Green finance and investment are highlighted as critical drivers of environmental transformation. The study examines the growing role of financial markets in advancing the transition to a green economy, particularly through the rise of green bonds, sustainable investments, and increased corporate responsibility. The financial sector's engagement is essential for funding and accelerating technological and industry shifts. Together, these areas provide a detailed blueprint for addressing environmental challenges through technological innovation and economic reform.

1.4.2.4. Governance, Policy, and Political Change

Transformational environmentalism necessitates significant changes in governance and policy at local, national, and global levels. The study explores the crucial role of environmental management and international agreements in shaping global environmental policies. Key agreements, such as the Paris Agreement and the Convention on Biological Diversity, serve as foundational frameworks guiding collective action to mitigate climate change, preserve ecosystems, and promote sustainability on a global scale.

At the local level, decentralized governance and community-led initiatives are highlighted as essential drivers of environmental transformation. The study underscores the importance of empowering local communities to implement sustainable practices, emphasizing that grassroots efforts can create meaningful change from the ground up. Local governance structures are vital for addressing region-specific environmental challenges, ensuring that solutions are contextually relevant and inclusive.

Additionally, the study examines the policy frameworks necessary for supporting transformational change. National and regional policies that promote green transitions—such as carbon pricing, subsidies for renewable energy, and regulations on sustainable land use—are critical in enabling countries to reduce emissions and embrace sustainable practices. These policies provide the structural support needed to incentivize green innovation, encourage responsible resource management, and align economic growth with environmental stewardship.

By addressing governance at multiple levels, the study illustrates the interconnected nature of policy-making. It emphasizes that coordinated action is required to achieve the large-scale environmental transformation to protect the planet for future generations.

1.4.2.5. Social Movements and Cultural Shifts

The study also delves into the significant role of social movements and cultural shifts in advancing transformational environmentalism. One key area of focus is grassroots activism and environmental movements. Social movements like Fridays for Future, Extinction Rebellion, and indigenous-led conservation efforts have raised awareness and catalyzed meaningful environmental change. These movements have mobilized communities globally, demanding action from governments and corporations to address climate change and environmental degradation.

The study further emphasizes the importance of a cultural shift in how societies value nature. It highlights the need to move away from an exploitative relationship with the environment toward one based on stewardship and ecological consciousness. This cultural transformation is essential for fostering a deeper connection to the natural world and promoting long-term sustainability.

Environmental education and public engagement are also critical elements explored in the study. It discusses the role of education in fostering environmental literacy, equipping individuals with the knowledge to make informed decisions about environmental issues. By encouraging citizen participation in environmental decision-making, education can empower people to take an active role in shaping policies and practices that protect the planet.

1.4.2.6. Social and Environmental Justice

The study further underscores the importance of integrating social justice into environmentalism. Environmental inequality is a key topic, highlighting how marginalized communities disproportionately bear the brunt of environmental degradation while often being excluded from the benefits of environmental progress.

Indigenous rights and knowledge are also critical components. The study examines the role of indigenous knowledge and practices in promoting sustainable environmental management and protecting biodiversity. Indigenous communities have long been stewards of the land, and their knowledge is invaluable in shaping conservation efforts.

The concept of just transitions is explored, ensuring that the shift to a sustainable economy is equitable. This involves ensuring that workers and communities dependent on carbon-intensive industries are not left behind and that opportunities are created for all as society moves toward a greener future.

1.4.2.7. Vision for a Sustainable Future

The study presents a vision for the future of environmentalism by exploring key pathways and strategies for a more sustainable world. One major focus is sustainability transitions, which examines how societies can shift toward sustainable production, consumption, and governance systems. These transitions involve rethinking current economic and environmental practices to create models prioritizing ecological balance, resource efficiency, and long-term environmental health.

The study also emphasizes the importance of global collaboration in addressing environmental challenges. International cooperation is critical to achieving sustainability goals, as environmental crises such as climate change, biodiversity loss, and pollution transcend national borders. By fostering partnerships and coordinated action, the global community can work collectively to implement solutions that benefit both people and the planet.

The study highlights the vital role of individuals and communities in driving transformational environmentalism. It explores how individual actions, community initiatives, and the efforts of civil society can contribute to sustainable change at every level—from local grassroots movements to global advocacy. This collective involvement is crucial for creating a more just, resilient, and environmentally conscious future for all.

The book seeks to provide a comprehensive and interdisciplinary approach to understanding and promoting systemic environmental change. Through its analysis of global environmental crises, theoretical frameworks, technological innovations, policy shifts, and social movements, the study aims to outline a clear pathway for achieving a sustainable future. Focusing on transformational change, the book calls for radically rethinking humanity's relationship with the natural world. It offers practical solutions for overcoming the barriers to achieving environmental sustainability and justice.

Chapter 2

Theoretical Foundations of Environmentalism

Environmentalism has evolved as a movement and philosophical stance over centuries, drawing from various intellectual traditions. Its theoretical foundations encompass diverse perspectives, including ecological science, moral philosophy, economics, and political theory. These frameworks have helped shape the environmental movement, guiding how humanity understands and interacts with the natural world. Below is an exploration of the key theoretical foundations that underpin environmentalism (Blanc, 2019; De Jonge, 2017; Haluza-DeLay, 2008).

2.1. The Key Theoretical Foundations That Underpin Environmentalism

2.1.1. Ecocentrism and Deep Ecology

Ecocentrism places intrinsic value on all living organisms and their natural environment, regardless of their utility to humans. This philosophy contrasts with anthropocentrism, which views nature primarily regarding its usefulness to human beings. Ecocentrism emphasizes the interconnectedness of all life forms and the importance of maintaining the health of entire ecosystems, not just individual species.

Deep Ecology, a concept introduced by Norwegian philosopher Arne Næss, extends this ecocentric view by advocating for a fundamental shift in humanity's relationship with nature. Deep ecology emphasizes the idea that all living beings have intrinsic value and that human activities should be aligned with the limits of the natural world. It challenges the anthropocentric view of environmentalism, arguing that the well-being of the Earth must take precedence over individual human needs and desires. The deep ecological movement calls for radical changes in societal values, consumption patterns, and economic structures to reduce humanity's ecological footprint (Kopnina, 2012; Sessions, 1987; Sessions, 1988).

Key principles of deep ecology include:

- Recognition of the intrinsic worth of all living beings.
- A focus on the complexity and interdependence of ecosystems.
- Advocacy for bioregionalism, where human societies are designed around the natural limits and capacities of the ecosystems they inhabit.

2.1.2. Anthropocentrism and Sustainable Development

While anthropocentrism centers ethical considerations on human needs and desires, it has also evolved to incorporate notions of sustainable development. This approach attempts to balance human progress with environmental protection by promoting resource use that does not compromise future generations' ability to meet their needs.

The concept of sustainable development gained global prominence with the Brundtland Report in 1987, which defined it as "development that meets the needs of the present without compromising the ability of future generations to meet their own needs." This view recognizes the interdependence of economic, social, and environmental factors and advocates for policies and practices that promote economic growth while conserving the environment (Speed, 2006; Adelman, 2018; Kopnina, 2020).

- *Weak Sustainability*: This version of sustainable development assumes that natural capital (environmental resources) can be substituted with human-made capital (technological innovations, infrastructure) if it depletes (Gutés, 1996).
- *Strong Sustainability*: This more ecocentric perspective argues that certain aspects of natural capital are irreplaceable and should be preserved for the well-being of future generations (Ott, 2003).

Sustainable development frameworks have shaped global environmental agreements, such as the Paris Agreement and the Sustainable Development Goals (SDGs), which integrate environmental concerns with social and economic priorities.

2.1.3. Environmental Ethics

Environmental ethics is a branch of philosophy that examines the moral relationship between humans and the natural environment. It challenges the traditional view of nature as a resource for human exploitation and promotes the idea that non-human entities—such as animals, plants, and ecosystems—have moral value (Palmer et al., 2014; Rolston, 1988; Elliot, 1995).

There are several schools of thought within environmental ethics:

- *Biocentrism*: This ethical view extends moral consideration to all living things, not just humans. It argues that all organisms have a right to exist and flourish, and human activities should not unduly harm other life forms (Baird, 2001).
- *Holism:* Popularized by philosopher Aldo Leopold in his Land Ethic, holism posits that ethical treatment should extend to individual organisms and entire ecosystems and landscapes. Leopold argued that human beings are part of a larger ecological community and that moral actions must promote the entire community's health, including the land, water, and species that inhabit it (Leopold, 1992).
- *Animal Rights and Welfare*: Philosophers such as Peter Singer and Tom Regan have advocated for animal rights, arguing that non-human animals should be given moral consideration, particularly their capacity to suffer. This ethical framework has influenced debates on biodiversity, wildlife conservation, and the humane treatment of animals.

2.1.4. Political Ecology and Eco-Marxism

- *Political ecology* examines the relationships between political, economic, and social structures and environmental issues. It critiques how power dynamics and inequalities shape human interactions with the environment. It emphasizes that environmental degradation is often linked to social injustice, unequal resource distribution, and exploitation by powerful interests (Cassegård, 2017).
- *Eco-Marxism*: This school of thought builds on Karl Marx's ideas, analyzing how capitalist modes of production drive environmental degradation. Eco-Marxism views environmental destruction as an

inherent consequence of capitalism, prioritizing economic growth, profit, and resource extraction over environmental sustainability. From this perspective, the solution requires a radical restructuring of global financial systems to prioritize the environment and social equity over profit and consumption (Barry, 1999).

- *Political ecology* also examines the role of globalization, colonialism, and indigenous rights in environmental issues. It advocates for equitable resource distribution and democratic participation in environmental governance, challenging the power of corporations and governments that exploit natural resources at the expense of local communities and ecosystems (Leonardi and Torre, 2022).

2.1.5. Systems Thinking and Resilience Theory

Environmental issues are complex and interconnected, making systems thinking a crucial theoretical approach. Systems thinking views the environment as an integrated whole, where various components interact dynamically, such as air, water, soil, plants, and animals. The goal is to understand the feedback loops and interactions between these components to manage ecosystems sustainably (Pisano, 2012; Wright et al., 2012; Curtin and Parker, 2014).

- *Resilience Theory*: Emerging from ecological science, resilience theory focuses on the ability of ecosystems and human societies to absorb shocks, adapt to change, and recover from disturbances. This theory is especially relevant in climate change, as it emphasizes the importance of building resilient systems that can withstand the increasing frequency of natural disasters, biodiversity loss, and other environmental stresses (Pisano, 2012).
- *Adaptive management* is a crucial concept within resilience theory. It involves flexible decision-making that adjusts to new information and changing circumstances rather than relying on rigid, top-down policies (Curtin and Parker, 2014). This approach is increasingly essential for managing complex, unpredictable environmental challenges.

2.1.6. Ecofeminism

Ecofeminism explores the connections between the exploitation of nature and the oppression of women, arguing that the domination of both stems from patriarchal and capitalist systems. Ecofeminists contend that the same structures that lead to environmental degradation perpetuate gender inequality (Shiva and Mies, 2014; Bhandari, 2024, 2024a).

- *Cultural Ecofeminism*: This strand of ecofeminism links the subjugation of women with the destruction of nature, arguing that patriarchal societies devalue both women's roles and the natural world (Gaard, 1993).
- *Social Ecofeminism*: Social ecofeminists emphasize the intersectionality of environmental degradation, gender oppression, class exploitation, and racism. They argue that addressing environmental issues requires a broader transformation of social structures that oppress marginalized groups (Gaard, 2017).

Ecofeminism calls for a more inclusive, nurturing, and cooperative approach to environmental stewardship, emphasizing care for people and the planet.

2.1.7. Conservation Biology and Biodiversity

Conservation biology is a scientific discipline that emerged in the 1980s with the goal of understanding and preserving biodiversity. It combines ecology, genetics, and social science elements to protect species, habitats, and ecosystems from extinction and degradation (Haila and Kouki, 1994; Olson et al., 2002; Hunter and Gibbs, 2006). The foundational principles of conservation biology include:

- *Species and Habitat Preservation*: Efforts to conserve species through protected areas, wildlife corridors, and biodiversity reserves.
- *Ecosystem Services*: Recognition that ecosystems provide essential services, such as clean water, pollination, and climate regulation, which are vital for human well-being.

- *Biodiversity Hotspots*: Identify areas of the world that are particularly rich in biodiversity and vulnerable to human impacts and focus conservation efforts on these regions.

Conservation biology has influenced international agreements, such as the Convention on Biological Diversity (CBD), which aims to halt biodiversity loss and protect ecosystems.

The theoretical foundations of environmentalism are diverse and multifaceted, drawing from ecological science, philosophy, economics, political theory, and social justice. These frameworks provide the intellectual tools necessary to understand environmental issues' complexity and guide efforts toward sustainable and just solutions. From the deep ecological commitment to intrinsic value and interconnectedness to the pragmatic focus on sustainable development, environmental ethics, and resilience, these theoretical approaches offer valuable insights for achieving a more harmonious relationship with the natural world.

2.1.8. Ecological Theories and Environmental Transformation

Ecological theories provide the scientific foundation for understanding the interactions between organisms and their environment. These theories are essential for framing how ecosystems function and respond to disturbances, and they underpin efforts to achieve environmental transformation (Mol, 2000; Bruckmeier, 2016; Brand and Wissen, 2017). As the world faces climate change, biodiversity loss, and ecosystem degradation, ecological theories are increasingly applied to design environmental sustainability, restoration, and transformation strategies.

2.1.8.1. Ecosystem Theory

Ecosystem theory emphasizes the interconnectedness of organisms and their environment, focusing on energy, matter, and nutrients flow through ecosystems. It views ecosystems as complex, adaptive systems where living organisms (biotic factors) and their physical environment (abiotic factors) interact dynamically (Müller, 1997; O'Neill and Reichle, 1979; Daily and Matson, 2008).

Key components of ecosystem theory:

- *Energy Flow:* Ecosystem theory highlights how energy flows from primary producers (plants and algae) to consumers (herbivores, carnivores) and decomposers, forming food webs. Energy transformations are governed by the laws of thermodynamics, where energy is transferred but not fully conserved, resulting in energy loss at each trophic level.
- *Nutrient Cycling*: Ecosystems recycle nutrients, such as nitrogen, carbon, and phosphorus, through biogeochemical cycles. These cycles sustain life by ensuring that essential elements are reused and redistributed.
- *Homeostasis and Stability*: Ecosystems maintain a state of equilibrium through feedback loops, where processes in another part counterbalance changes in one part of the system. However, human activities, such as deforestation, pollution, and climate change, can disrupt this balance, leading to ecosystem collapse.

Ecosystem theory has been crucial in shaping environmental transformation initiatives, particularly in:

- *Ecosystem-Based Management*: This approach manages natural resources by considering the entire ecosystem rather than focusing on a single species or resources. It emphasizes the need for ecosystem health and resilience to support long-term sustainability.
- *Ecological Restoration*: Ecosystem theory informs restoration efforts, which aim to rehabilitate degraded ecosystems by reintroducing native species, restoring natural processes, and removing invasive species. Restoration seeks to return ecosystems to a state of balance, increasing biodiversity and resilience to climate change.

2.1.8.2. Succession Theory

Succession theory describes how ecosystems change over time, particularly after disturbances such as fires, floods, or human activities. The theory identifies two main types of ecological succession:

Primary Succession occurs in lifeless areas, such as newly formed volcanic islands or regions exposed after a glacier retreats. Over time, pioneer species (lichens, mosses) colonize the area, gradually developing more complex ecosystems.

Secondary Succession occurs in areas where an ecosystem has been disturbed but not destroyed, such as after a forest fire or agricultural land abandonment. Ecosystems regenerate through regrowth stages, eventually reaching a stable and self-sustaining climax community.

Succession theory is important for environmental transformation because it highlights the following (McIntosh, 1981; Glenn-Lewin et al., 1992; Peet and Christensen, 1980).

- *Natural Regeneration:* In some cases, ecosystems can recover naturally from disturbances if human pressures are reduced or eliminated. This understanding is central to rewilding projects and other forms of passive restoration, where human intervention is minimized.
- *Adaptive Management:* Succession theory emphasizes that ecosystems are dynamic and constantly evolving. Environmental transformation initiatives must be flexible and adaptive, responding to ecological changes over time rather than seeking to maintain a static, unchanging environment.

2.1.8.3. Resilience Theory

Resilience theory focuses on the capacity of ecosystems and human societies to absorb disturbances, adapt to change, and recover from shocks without collapsing into a different state. It defines resilience as the ability to persist in adversity while maintaining essential functions and structures (Van Breda, 2001; Masten, 2018; Carlson et al., 2012). The Key concepts in resilience theory:

- *Thresholds and Tipping Points*: Ecosystems can tolerate certain stress levels before they reach a tipping point, shifting into a new state that may be less productive or less biodiverse. For example, an overexploited forest may change to a grassland state if it passes a critical threshold.
- *Panarchy:* This concept describes the interconnectedness of systems at different scales (local, regional, global). Changes at one scale can have cascading effects on other scales, such as how local deforestation contributes to global climate change.
- *Adaptive Cycles*: Resilience theory suggests that ecosystems go through cycles of growth, accumulation, collapse, and renewal. These

cycles are natural and necessary for long-term system health, but human interventions often disrupt them.

In environmental transformation, resilience theory is applied to:

- *Climate Adaptation*: Resilience theory informs strategies for helping ecosystems and human societies adapt to the impacts of climate change. This includes designing cities, agricultural systems, and natural reserves that withstand extreme weather, rising sea levels, and other climate-related disturbances.
- *Disaster Risk Reduction*: Understanding ecosystem resilience helps manage natural hazards such as floods, hurricanes, and wildfires. Healthy ecosystems, such as wetlands and forests, can buffer against these hazards, reducing their impact on human communities.
- *Sustainable Development*: Resilience theory calls for economic and development models prioritizing long-term sustainability over short-term gains. It advocates for policies and practices that enhance ecological resilience, ensuring that ecosystems can continue to provide essential services like clean water, food, and air.

2.1.8.4. Island Biogeography and Conservation Biology

Island biogeography theory, developed by ecologists Robert MacArthur and E. O. Wilson, explains how species richness on islands is determined by the size of the island and its distance from the mainland. The theory has been expanded to apply to isolated habitats, such as nature reserves, fragmented forests, or mountain ranges (Whittaker and Fernández-Palacios, 2007; Kent, 1987; Matthews and Triantis, 2021). Key concepts of island biogeography:

- *Species-Area Relationship*: Larger islands (or habitats) tend to support more species due to greater habitat diversity and lower extinction rates.
- *Isolation and Immigration:* Islands (or habitats) that are closer to the mainland (or other habitats) receive more new species through immigration, which increases species richness.

Island biogeography theory has profound implications for environmental transformation, particularly in the fields of conservation and biodiversity management:

- *Habitat Fragmentation*: Human activities fragment natural habitats, creating isolated "islands" of biodiversity. The theory helps explain how species loss occurs and guides conservation efforts to connect fragmented habitats with wildlife corridors.
- *Protected Area Design*: The theory informs the design of nature reserves, emphasizing the importance of size and connectivity to ensure that ecosystems remain viable and species can migrate between habitats.

2.1.8.5. Niche Theory and Biodiversity

Niche theory explores each species' role in an ecosystem, including how it interacts with other species and its environment. Every species occupies a unique ecological niche, including its habitat, role in the food web, and the conditions it needs to survive and reproduce (Elena and Gleb, 2013; Barton et al., 2020; Gatti, 2011). The key aspects of niche theory:

- *Competitive Exclusion Principle*: No two species can occupy the same niche indefinitely because they will compete for the same resources, eventually causing one species to be extinct. This principle explains why biodiversity is essential for ecosystem health, as diverse species fill different ecological roles.
- *Niche Differentiation*: Species can coexist by occupying different niches, such as birds that feed on different parts of the same tree or animals that forage at different times.

Niche theory is critical for understanding the impacts of environmental transformation, particularly in the context of:

- *Biodiversity Conservation*: Protecting biodiversity requires maintaining a diversity of niches within ecosystems. If niches are lost due to habitat destruction or climate change, species that rely on those niches may go extinct.
- *Invasive Species Management*: Invasive species can outcompete native species by occupying their niche, leading to the decline or extinction of native biodiversity. Understanding niche theory helps design strategies to control invasive species and restore native ecosystems.

2.1.8.6. Gaia Hypothesis

The Gaia hypothesis, proposed by James Lovelock, posits that the Earth functions as a self-regulating system. Living organisms and their inorganic surroundings form a complex, interacting system that maintains conditions conducive to life. According to this theory, life on Earth influences the atmosphere, oceans, and land in ways that stabilize the planet's environment (Kirchner, 1989; Free and Barton, 2007; Kirchner, 2002). The key insights from the Gaia hypothesis:

- *Homeostasis and Self-Regulation*: The Earth system has feedback loops that regulate temperature, ocean salinity, atmospheric composition, and other conditions essential for life. For example, plants regulate the amount of carbon dioxide in the atmosphere through photosynthesis.
- *Global Ecological Balance*: The Gaia hypothesis suggests that the Earth's biosphere is a single, self-sustaining entity. Human activities, such as burning fossil fuels and deforestation, disrupt this balance, leading to global climate change and biodiversity loss.

In environmental transformation, the Gaia hypothesis encourages:

- *Systems Thinking*: It reinforces the need to view the Earth as an interconnected system where changes in one part of the planet (e.g., deforestation in the Amazon) can have far-reaching impacts (e.g., global climate patterns).
- *Sustainable Living*: The hypothesis advocates for lifestyles and policies that align with the Earth's natural regulatory processes, emphasizing the importance of reducing pollution, conserving ecosystems, and living within planetary boundaries.

2.1.8.7. Ecological Economics and the Carrying Capacity Concept

Ecological economics integrates ecological principles with economic theory to address the limits of economic growth to the Earth's finite resources. It challenges the traditional economic focus on infinite growth, arguing that economies must function within the carrying capacity of ecosystems (Sagoff, 1995; Munda, 1997).

Carrying Capacity refers to the maximum population size or level of resource consumption that an ecosystem can sustain indefinitely without degrading. Ecological economics highlights that exceeding carrying capacity leads to resource depletion, environmental degradation, and ecological collapse (Zhao et al., 2009).

Environmental transformation guided by ecological economics emphasizes:

- *Sustainable Resource Use*: Policies must ensure that resource consumption does not exceed the regenerative capacity of ecosystems. This includes promoting renewable energy, sustainable agriculture, and circular economy practices that minimize waste and resource use.
- *Economic Models Aligned with Nature*: Ecological economics advocates shifting from GDP-based growth models to those that value ecosystem services, human well-being, and sustainability over pure economic output.

Ecological theories are central to understanding how ecosystems function, change, and respond to human activities. They offer frameworks for conceptualizing environmental transformation, from ecosystem-based management and restoration to resilience and sustainable resource use (Sagoff, 1995; Munda, 1997; Zhao et al., 2009). By applying these theories, societies can design strategies for addressing pressing environmental challenges, such as climate change, biodiversity loss, and habitat destruction, and move towards a more sustainable and resilient future.

2.2. Integral Theory and Holistic Environmentalism

Integral Theory, pioneered by philosopher Ken Wilber, offers a framework for understanding complex systems and integrating multiple perspectives into a comprehensive, holistic worldview. It aims to synthesize various fields of knowledge, from science and philosophy to spirituality and social sciences, into a cohesive approach that can address the multifaceted nature of environmental issues. When applied to environmentalism, integrative theory leads to holistic environmentalism, which views environmental issues not merely through a singular lens but as complex, interconnected challenges that require integrated solutions (Esbjörn-Hargens, 2005; Esbjörn-Hargens and Zimmerman, 2009; Arkell, 2018).

2.2.1. The Four Quadrants of Integral Theory in Environmentalism

Integral Theory's Four Quadrants model is a central tool for organizing different dimensions of experience and reality. Each quadrant represents a different aspect of life, which must be considered in any comprehensive solution to environmental problems (Esbjörn-Hargens, 2005; Zimmerman, 2005; Esbjörn-Hargens and Zimmerman, 2009). The four quadrants are:

- *Upper Left (UL): Individual Interior (Intentional/Subjective):* This quadrant concerns individuals' inner experiences, including their values, beliefs, attitudes, and consciousness. Environmentalism represents the personal motivations and ethics that drive individuals to act on behalf of the environment.
- *Upper Right (UR): Individual Exterior (Behavioral/Objective):* This quadrant focuses on individuals' observable actions and behaviors. Environmentalism includes individual behaviors such as energy consumption, recycling, and lifestyle choices directly impacting the environment.
- *Lower Left (LL): Collective Interior (Cultural/Inter-subjective):* This quadrant addresses shared values, culture, and worldviews. Cultural values, collective beliefs, and societal norms deeply shape environmental movements. Cultural environmentalism examines how environmental consciousness spreads through art, media, and education, influencing public discourse and collective behavior.
- *Lower Right (LR): Collective Exterior (Systems/Objective)*: This quadrant pertains to social systems, institutions, and external structures. Environmentalism refers to the political, economic, and technological systems influencing environmental policy, corporate behavior, and industrial impacts.

Applying the Four Quadrants to Environmental Transformation:

- *UL (Individual Interior):* We must go beyond mere awareness to address environmental crises like climate change and biodiversity loss. We need a deeper ecological consciousness. This transformational environmentalism at the personal level fosters a sense of personal responsibility, ecological identity, and connection to nature through education, mindfulness, and ethical living.

- *UR (Individual Exterior):* Personal environmental action must accompany awareness. This includes adopting sustainable behaviors like reducing energy usage, choosing sustainable products, and minimizing waste. It also involves individuals advocating for environmental causes, contributing to policy changes, or supporting green technologies.
- *LL (Collective Interior):* Environmentalism must address cultural worldviews, shifting societal norms towards sustainability and ecological integrity. This involves changing public attitudes through environmental education, storytelling, community initiatives, and cultural movements that value harmony with nature. Indigenous worldviews, which often embody a deep respect for nature, can play an important role in this transformation.
- *LR (Collective Exterior):* Systemic change is crucial to achieving environmental sustainability. This quadrant calls for transforming political institutions, economic systems, and technological infrastructures to support sustainable practices. Environmental regulations, carbon pricing, renewable energy innovations, and sustainable urban planning are examples of how systemic structures can be reformed to promote environmental health.

2.2.2. Holistic Environmentalism: Integrating Multiple Perspectives

Holistic Environmentalism, rooted in Integral Theory, recognizes that environmental problems arise from multiple interacting factors—individual actions, societal values, technological innovations, and political systems—all of which must be addressed simultaneously. This approach encourages an inclusive, multidimensional strategy for environmental action that integrates ecological, social, economic, and spiritual considerations (Patten, 2014; Grizzle, 1994; Inoue and Franchini, 2020).

2.2.2.1. An Ecological Perspective

Environmentalism focuses on the planet's ecological health—preserving ecosystems, biodiversity, and natural resources. Holistic environmentalism goes beyond traditional conservation efforts by recognizing that ecological solutions require addressing root causes like consumption patterns, industrial practices, and land use planning (McLaren and Hawe, 2005; Gitterman, 2009; Kramsch, 2008). This includes:

- *Ecosystem-Based Management*: Integrating ecological knowledge into land and resource use decisions to maintain ecosystem integrity.
- *Sustainable Agriculture and Food Systems*: Promoting agricultural practices that protect soil health, water resources, and biodiversity while ensuring food security.

2.2.2.2. A Social and Cultural Perspective

Holistic environmentalism acknowledges that environmental issues are also social and cultural issues. It addresses environmental justice, recognizing that marginalized communities often face environmental degradation through pollution, climate impacts, or lack of access to clean resources (Pharino, 2019; Bhandari, 2023bc, 2024). Key approaches include:

- *Environmental Justice:* Advocating for fair distribution of environmental benefits and burdens and ensuring that vulnerable populations have a voice in environmental decision-making.
- *Cultural Sustainability*: Promoting cultural practices that support environmental stewardship, including indigenous and traditional knowledge systems that offer valuable insights into sustainable living.

2.2.2.3. An Economic Perspective

Economic systems play a crucial role in shaping environmental outcomes. Holistic environmentalism calls for transforming current economic models to prioritize sustainability over endless growth (Tol, 2009, 2014; Adams, 1989; Bhandari, 2022, 2024). This includes:

- *Green Economy*: Shifting towards economic systems that promote sustainable resource use, renewable energy, and environmentally friendly industries.
- *Circular Economy*: Encouraging a system where waste is minimized and resources are reused and recycled, reducing pressure on natural ecosystems.

2.2.2.4. A Spiritual and Ethical Perspective

Environmentalism is about scientific and practical solutions and is deeply connected to human values, ethics, and spirituality. Holistic environmentalism encourages a spiritual connection to nature, emphasizing the intrinsic value of all living beings and ecosystems (Smith and Pulver, 2009; White, 2010; Taylor, 2001). This perspective fosters:

- *Ecological Ethics*: Emphasizing moral obligations to protect the environment, including non-human species, for current and future generations.
- *Spiritual Connection to Nature*: Recognizing the sacredness of nature and fostering deep personal connections to the earth through practices such as eco-spirituality, indigenous traditions, and mindfulness in nature.

2.2.3. Applying Integral Environmentalism in Practice

Holistic environmentalism, informed by Integral Theory, provides practical pathways for individuals, communities, and societies to engage in environmental transformation (Fisher, 2019; Banerjee, 2017; Ferrara, 2019). This requires coordinated action across all quadrants:

- *Personal and Collective Action*: Encouraging individual behavior changes (recycling, reducing carbon footprints) and collective cultural shifts (changing social norms around consumption).
- *Policy and Systemic Change*: Advocating for institutional changes such as stronger environmental regulations, economic incentives for sustainability, and corporate responsibility to ensure that entire systems align with ecological values.
- *Education and Awareness*: Fostering environmental education that integrates scientific knowledge with cultural understanding, spiritual wisdom, and ethical responsibility. Programs in schools, communities, and workplaces can encourage more holistic ways of thinking and acting regarding environmental issues.
- *Technology and Innovation*: Supporting technological advancements that contribute to environmental sustainability, such as clean energy, sustainable agriculture, and resource-efficient technologies, while critically examining their social and environmental impacts.

2.2.4. Challenges and Opportunities in Holistic Environmentalism

While Holistic Environmentalism offers a comprehensive framework for addressing environmental challenges, there are obstacles to its full implementation (Thiele, 1999; Patten, 2014; Marcinkowski, 2009).

- *Complexity and Scale:* Environmental issues' interconnected nature can make them difficult to address. Holistic environmentalism requires collaboration across multiple sectors, including governments, industries, civil society, and local communities.
- *Resistance to Change*: Shifting cultural and institutional paradigms can be met with resistance from entrenched interests, whether in fossil fuel industries, political systems, or consumer behaviors.
- *Integration of Diverse Worldviews*: Holistic environmentalism aims to integrate diverse perspectives, from scientific to spiritual, but achieving this integration requires overcoming divisions between different disciplines and worldviews.

Despite these challenges, Holistic Environmentalism offers a vision of transformation that is both inclusive and far-reaching. Recognizing ecological, social, economic, and spiritual interdependence paves the way for a more sustainable and equitable future. It calls for a global shift in consciousness and practice, rooted in the understanding that humanity's well-being is inseparable from the planet's well-being.

Integral Theory and Holistic Environmentalism provide a robust framework for addressing environmental issues' complex, interconnected nature. By integrating personal, cultural, behavioral, and systemic dimensions, this approach emphasizes the need for coordinated, multi-dimensional action. Holistic Environmentalism goes beyond traditional conservation efforts, offering a vision of environmental transformation that encompasses social justice, economic reform, spiritual connection, and ecological sustainability. This comprehensive lens allows societies to move towards a more balanced and harmonious relationship with the natural world.

2.3. The Capabilities Approach in an Environmental Context

The Capabilities Approach, developed by Amartya Sen and Martha Nussbaum, is a framework for evaluating human well-being based on the opportunities and freedoms individuals have to live the lives they value. Rather than focusing solely on material wealth or economic output, the approach emphasizes the capabilities or fundamental freedoms people have to achieve their potential. In environmentalism, the Capabilities Approach can provide a powerful tool for understanding the relationship between human development, environmental sustainability, and justice (Eichsteller et al., 2022; Lessmann and Rauschmayer, 2013; Roesler, 2011).

2.3.1. Core Concepts of the Capabilities Approach

At its core, the Capabilities Approach focuses on two key concepts. The first is functioning, which refers to the various things a person values being or doing, such as being healthy, having access to education, or participating in a community. These functions represent the essential aspects of a person's well-being and quality of life (Schokkaert, 2008; Gasper, 2017; Robeyns, 2006).

The second concept is capabilities, the freedoms or opportunities a person has to achieve these valued functionings. Both individual circumstances and the broader social, economic, and environmental context shape capabilities. In other words, capabilities determine what a person can do and what they are truly free to do given their situation.

In an environmental context, the Capabilities Approach encourages a shift in focus beyond the mere availability of resources like water or clean air. It emphasizes whether individuals truly have access to these resources and whether they can live sustainably. This approach highlights the importance of ensuring that environmental benefits and opportunities are equitably distributed, enabling people to thrive in harmony with nature.

2.3.2. Environmental Justice and the Capabilities Approach

The Capabilities Approach provides a strong foundation for addressing environmental justice because it highlights the importance of equitable access to environmental resources and the ability to enjoy a healthy, sustainable environment. Environmental justice goes beyond distributing environmental goods (like clean water or green spaces). It seeks to ensure that all individuals, regardless of socioeconomic status, race, or location, have the capabilities to live in a healthy and safe environment (Day, 2017; Edwards et al., 2016; Holland, 2008).

- *Capabilities and Environmental Inequality*: The approach is beneficial for analyzing how environmental degradation and unequal access to resources disproportionately affect marginalized communities. For example, a wealthy individual may be able to move away from polluted areas or buy bottled water. At the same time, a poorer person may lack these options, diminishing their ability to lead a healthy life.
- *Agency and Empowerment*: In environmental terms, the Capabilities Approach also stresses the importance of agency—people's ability to participate in decision-making processes that affect their environment. Environmental policies that empower communities to shape their surroundings (such as community-based resource management) are aligned with the Capabilities Approach's emphasis on freedom and participation.

2.3.3. Sustainable Development Through the Capabilities Lens

The Capabilities Approach is closely aligned with sustainable development principles, which balance economic growth, social inclusion, and environmental sustainability. A capabilities-focused approach to environmental sustainability emphasizes (Wolbring and Burke, 2013; Stenn, 2016; Behnam et al., 2018; Bhardwaj et al., 2021).

- *Human Well-being and Ecological Integrity*: Policies that promote environmental sustainability must ensure that they also enhance human well-being. For instance, renewable energy projects can reduce carbon emissions and improve local health by reducing air pollution and expanding people's capabilities.
- *Long-term Environmental Justice*: Future generations must also have the capabilities to enjoy a healthy environment. A capabilities-based approach to sustainability promotes intergenerational justice, ensuring that today's environmental decisions do not erode the future ability of people to lead flourishing lives.

- *Inclusive Development*: The Capabilities Approach emphasizes that sustainable development should focus on aggregate economic growth and whether it provides meaningful improvements in people's lives. Projects should ensure that economic development aligns with expanding individuals' real opportunities without sacrificing environmental integrity.

2.3.4. Ecological Capabilities

The concept of ecological capabilities extends the framework of the Capabilities Approach to include the environment as a fundamental capability necessary for human well-being. Without access to clean air, water, fertile soil, and a stable climate, individuals cannot fully develop other capabilities, such as health, education, and social participation (Holland, 2008; Gross and Wilson, 2020; Gabler et al., 2015).

2.3.4.1. Basic Environmental Capabilities

Basic environmental capabilities can be understood as the essential ecological goods and services that support life and human flourishing, such as:

- *Clean Water:* Access to clean drinking water is a fundamental environmental capability supporting health and hygiene.
- *Clean Air:* The ability to breathe unpolluted air directly impacts health and well-being.
- *Sufficient Food*: Ecological sustainability in agriculture ensures that food systems are resilient and capable of providing adequate nutrition for all.

2.3.4.2. Capabilities to Participate in Environmental Decision-Making

Environmental policies should also enable individuals and communities to play a meaningful role in shaping the environmental conditions affecting their lives. This includes the ability to:

- Participate in environmental governance (e.g., local resource management and conservation efforts).
- Access information about environmental hazards and their impacts on health and well-being.

- Advocate for environmental protections that enhance local and global ecological resilience.

2.3.5. The Capabilities Approach and Climate Change

Climate change presents a significant challenge to environmental justice, and the Capabilities Approach offers valuable insights into how climate policies can be designed to address human vulnerabilities and promote fairness (Roy and Venema, 2002; Eichsteller et al., 2022; Peeters et al., 2015; Dalziel et al., 2018).

- *Vulnerability to Climate Change:* The impacts of climate change are disproportionately felt by the most vulnerable populations, who often have the fewest capabilities to adapt. This includes low-income communities, indigenous peoples, and residents of small island nations or regions prone to extreme weather events. A capabilities-based approach emphasizes expanding adaptive capacities in these communities by providing resources, education, and technology to enhance their resilience.
- *Climate Justice*: From a capabilities perspective, climate justice is about ensuring that people have the capabilities to live healthy, productive lives even in the face of climate change. Policies must prioritize expanding the resilience and adaptive capacities of those at risk, ensuring they are not further disadvantaged by environmental policies or global agreements.
- *Intergenerational Climate Justice*: The Capabilities Approach supports the idea that future generations should have the same capabilities as current generations to enjoy a stable, healthy environment. This underscores the importance of mitigating climate change by reducing emissions and preserving ecosystems for the benefit of future generations.

2.3.6. Policy Implications of the Capabilities Approach in Environmentalism

The Capabilities Approach offers several practical policy implications for environmental governance and sustainability:

- *Inclusive and Participatory Environmental Governance*: Policies should empower individuals and communities to participate in environmental decision-making. This includes increasing access to information, supporting grassroots initiatives, and creating channels for marginalized communities to influence policies that affect their environment.
- *Addressing Environmental Inequalities*: Environmental policies should reduce inequalities in access to clean air, water, and natural resources. This may include targeted measures to address pollution in low-income areas, enforce stricter environmental regulations in marginalized communities, or provide clean energy solutions to underserved regions.
- *Sustainable Resource Use*: Environmental sustainability policies should balance using natural resources with the need to preserve them for future generations. This includes fostering renewable energy, promoting sustainable agricultural practices, and encouraging a circular economy that minimizes waste and resource depletion.
- *Human Development and Ecological Protection*: The Capabilities Approach calls for environmental policies prioritizing human development and ecological health. For example, economic growth strategies should be designed to avoid environmental degradation and promote sustainable livelihoods, especially in vulnerable communities.

The Capabilities Approach, focusing on expanding human freedoms and opportunities, offers a valuable framework for addressing the complex relationship between human well-being and environmental sustainability. In an environmental context, it emphasizes the importance of equitable access to natural resources, the capability to participate in environmental governance, and the need for climate justice for vulnerable populations (Reed, 2013; Anand, 2014; Lessmann and Rauschmayer, 2014). By prioritizing human development alongside ecological protection, the Capabilities Approach provides a holistic, justice-oriented pathway for achieving sustainable development and environmental resilience.

2.4. Systems Thinking in Environmental Transformation

Systems Thinking is an approach to problem-solving and understanding that emphasizes the interconnections and interactions within complex systems. In the context of environmental transformation, it provides a comprehensive framework for analyzing how different components of ecological, social, and economic systems interact and influence one another. This perspective is crucial for addressing environmental issues, which often involve multiple interconnected factors that cannot be effectively understood or managed through linear or isolated approaches (Voulvoulis et al., 2022; Fath, 2014; Herrmann et al., 2021).

2.4.1. Core Principles of Systems Thinking

Systems Thinking is built upon several core principles that guide the understanding and management of complex systems. The first principle is interconnectedness, which asserts that every component within a system is linked, and a change in one part can influence other parts. This principle underscores the necessity of considering the entire system when making decisions rather than focusing on isolated components. Recognizing these connections helps in predicting outcomes and avoiding unintended consequences (Voulvoulis et al., 2022; Fath, 2014; Herrmann et al., 2021).

Another fundamental principle is adopting a holistic view. Systems Thinking promotes the idea of understanding the entire system as a whole rather than merely analyzing individual parts. This approach emphasizes looking at the broader context and recognizing how various elements within a system interact and influence one another. By doing so, it allows for a more comprehensive perspective that captures the complexity of real-world systems.

Feedback loops are also central to Systems Thinking. These loops can either reinforce changes (positive feedback) or counteract them (negative feedback), thereby influencing the system's behavior over time. Identifying and understanding these feedback mechanisms is crucial for predicting how systems respond to interventions, enabling more effective management and planning.

Systems Thinking acknowledges the complexity and emergence inherent in systems. Complex systems exhibit emergent properties—characteristics that arise from the interactions between components, which cannot be predicted by examining the behavior of individual parts alone. This principle highlights the dynamic and sometimes unpredictable nature of systems, stressing the importance of adapting strategies based on the system's evolving behavior. Together, these principles form the foundation for analyzing and addressing intricate challenges in various fields, from environmental management to organizational development.

2.4.2. Applying Systems Thinking to Environmental Transformation

In environmental transformation, Systems Thinking helps address the complexity and interdependencies of environmental issues. It provides a framework for understanding how various factors contribute to environmental problems and how interventions can lead to unintended consequences (Voulvoulis et al., 2022; Wan Rosely and Voulvoulis, 2023; Wilkinson et al., 2018; Orr, 2014). Key applications include:

2.4.2.1. Understanding Ecological Systems

- *Ecosystem Dynamics:* Systems Thinking helps in understanding the dynamics of ecosystems, including nutrient cycling, energy flow, and species interactions. For example, a decline in one species can affect the entire food web and cascade effects throughout the ecosystem.
- *Resilience and Adaptation*: It emphasizes the concept of resilience—how ecosystems can absorb disturbances and adapt to changes. This includes studying how ecosystems recover from events like wildfires or floods and how they adapt to gradual changes like climate change.

2.4.2.2. Addressing Human-Environment Interactions

- *Resource Management*: Effective management of natural resources requires understanding the complex interactions between human activities and environmental systems. For instance, water management must consider not only water availability but also the impacts of consumption, pollution, and climate change on water cycles.

- *Sustainability and Development*: Systems Thinking supports sustainable development by examining the relationships between economic growth, social equity, and environmental health. It helps design development strategies that balance these dimensions and avoid negative impacts on ecosystems.

2.4.2.3. Policy Design and Implementation

- *Integrated Approaches*: Policies addressing environmental issues must integrate multiple factors, including economic incentives, regulatory frameworks, and community engagement. Systems Thinking supports the development of policies that consider the interconnected nature of environmental challenges and promote coordinated actions across sectors.
- *Scenario Planning*: Systems thinking aids in anticipating future challenges and designing adaptive strategies by exploring different scenarios and their potential impacts on systems. For example, scenario planning for climate change can help identify potential risks and develop strategies to mitigate or adapt to those risks.

2.4.3. Tools and Methods in Systems Thinking

Several tools and methods are commonly used in Systems Thinking to analyze and address environmental issues:

2.4.3.1. Systems Mapping

- *Causal Loop diagrams* visualize system feedback loops and interactions. They help understand how changes in one component can influence others and identify potential leverage points for intervention.
- *Stock and Flow Diagrams*: These diagrams illustrate a system's accumulation and movement of resources or entities. They are helpful in analyzing resource management, such as tracking water usage and its impacts on availability.

2.4.3.2. Modeling and Simulation

- *System Dynamics Models*: These models simulate the behavior of complex systems over time, allowing for the exploration of different scenarios and the impact of various interventions. For example, a system dynamics model can simulate the effects of varying conservation strategies on biodiversity and ecosystem services.
- *Agent-Based Models:* These models simulate interactions between individual agents (e.g., people, organizations) within a system to understand how their behaviors affect the system as a whole. They help study social dynamics and the impact of individual decisions on environmental outcomes.

Systems thinking encourages including multiple perspectives and stakeholder inputs in the analysis and decision-making process. Understanding different stakeholders' interests, influences, and interactions helps design more effective and inclusive solutions (Wilkinson et al., 2018; Cabrera and Cabrera, 2023; Salmon et al., 2022).

2.4.4. Challenges and Opportunities in Systems Thinking

While Systems Thinking provides valuable insights for environmental transformation, its application comes with several challenges. One significant challenge is managing complexity. Environmental systems are inherently intricate, with numerous interdependent components. Effectively addressing this complexity requires specialized tools and methods to capture and analyze the intricate relationships within the system. Navigating this complexity is crucial to developing informed strategies for meaningful environmental solutions (Basile and Caputo, 2017; Trochim et al., 2006; Pazicni and Flynn, 2019).

Another challenge lies in fostering interdisciplinary collaboration. Environmental issues often span multiple fields, requiring the integration of knowledge from disciplines such as ecology, economics, sociology, and political science. Systems Thinking encourages interdisciplinary approaches, but achieving effective collaboration can be difficult due to differences in terminology, methodologies, and perspectives. Bridging these gaps requires deliberate effort to create a shared understanding among diverse experts and stakeholders.

Data and uncertainty also pose challenges in the application of Systems Thinking. The approach relies heavily on models and simulations to predict system behavior, but uncertainties and gaps in available data can limit these models. In the context of environmental transformation, robust data collection and analysis methods are crucial to minimize uncertainty. Improving data quality and addressing limitations is essential for making sound, evidence-based decisions when facing environmental challenges.

Despite these hurdles, Systems Thinking remains a powerful tool for understanding and managing the complexities of environmental transformation, provided these challenges are actively addressed.

2.4.5. Case Studies and Examples

Several case studies illustrate the application of Systems Thinking in environmental transformation:

2.4.5.1. Urban Sustainability

Green infrastructure represents a prime example of how Systems Thinking can be applied to the design and management of urban environments by integrating natural systems into cities. Systems Thinking enables planners and policymakers to see urban spaces not just as static areas, but as dynamic systems that interact with the natural environment. This perspective is crucial for designing green infrastructure projects that simultaneously address multiple environmental, social, and economic goals (Maani, 2016; Nguyen et al., 2023; Sherwood, 2011).

One application of Systems Thinking in green infrastructure is the creation of green roofs. These vegetated rooftops serve as mini ecosystems that help absorb rainwater, reduce stormwater runoff, and improve urban water management. In addition, green roofs act as insulators, lowering energy consumption in buildings by keeping them cooler in the summer and warmer in the winter. By considering the interconnected benefits to the environment and urban living conditions, green roofs help mitigate the heat island effect. In this phenomenon, human activities make urban areas significantly warmer than their rural surroundings.

Similarly, developing urban parks and green spaces illustrates how Systems Thinking can foster multifunctional spaces that offer environmental and social benefits. Urban parks enhance biodiversity by providing habitats for various species and improving air quality by filtering pollutants and producing oxygen. From a social perspective, these spaces offer communities recreation, leisure, and social interaction, promoting mental and physical well-being. These multiple functions are interconnected and contribute to cities' overall sustainability and livability.

By integrating natural elements such as vegetation, water bodies, and permeable surfaces into urban design, green infrastructure also helps cities become more resilient to climate change. Systems Thinking highlights how these green projects can work together to reduce the impacts of extreme weather events, such as floods and heat waves (Huang et al., 2015; Verma and Raghubanshi, 2018; Alberti, 1996). The holistic approach of Systems Thinking ensures that green infrastructure serves as an aesthetic or recreational feature and contributes to long-term environmental sustainability and resilience.

2.4.5.2. Ecosystem-Based Management

Systems Thinking has been applied to marine conservation efforts, such as managing fisheries and protected areas. By considering the interactions between fish populations, fishing practices, and marine ecosystems, policymakers can develop strategies that support sustainable fishing practices and protect marine biodiversity (McLeod and Leslie, 2009; Arkema et al., 2006; Long et al., 2015).

2.4.5.3. Climate Change Adaptation

- *Community Resilience:* Systems Thinking supports climate change adaptation efforts by analyzing how communities can build resilience to climate impacts. For example, integrating flood risk management with land-use planning and community engagement helps create adaptive strategies that protect vulnerable populations and enhance overall resilience.

Systems Thinking provides a powerful framework for understanding and addressing the complexity of environmental transformation. By emphasizing interconnectedness, holistic analysis, and feedback loops, it supports the development of integrated and adaptive solutions to environmental challenges (Biesbroek et al., 2013; Franks and Hoffmann, 2012; Fankhauser, 2017). Applying Systems Thinking to environmental issues helps design policies and strategies considering the full range of interactions within ecological, social, and economic systems, ultimately contributing to more sustainable and resilient environmental outcomes.

Chapter 3

Drivers of Environmental Transformation

Environmental transformation refers to significant and often fundamental changes in ecological systems, processes, or conditions. These transformations can be driven by various factors—both natural and human-induced—that interact in complex ways to shape the state of the environment. Understanding these drivers is crucial for designing effective sustainable development and environmental management strategies (Shkaruba et al., 2021; Preisendörfer, 2017; Linnér and Wibeck, 2021).

3.1. Natural Drivers of Environmental Transformation

Natural drivers refer to factors originating from natural processes and phenomena that can lead to significant environmental changes over various timescales. These drivers operate through various mechanisms, influencing climate, geological formations, and biological processes (Pichler et al., 2017; Thaler et al., 2019; Linnér and Wibeck, 2019).

One major natural driver is climate variability and change. Natural climate variability, such as El Niño and La Niña, leads to temporary changes in global weather patterns and ocean currents, significantly affecting regional climates. These variations can alter temperature, precipitation levels, and the frequency of extreme weather events. In addition to short-term variability, long-term climate change has historically been driven by natural factors such as variations in Earth's orbit, solar radiation fluctuations, and volcanic activity. These processes have shaped ecosystems and climate over geological timescales. However, the current rate of climate change is vastly accelerated by human activities, particularly the emission of greenhouse gases.

Geological processes also play a critical role in driving environmental changes. The movement of tectonic plates can trigger volcanic eruptions, earthquakes, and the formation of mountain ranges, all reshaping landscapes and influencing ecosystems and climate patterns. Erosion and sedimentation, two other key geological processes, continuously shape the Earth's surface by transporting and depositing sediments, impacting soil formation, river patterns, and coastal dynamics. These natural processes contribute to the long-term evolution of physical landscapes and ecological habitats.

Biological and ecological processes are equally important natural drivers. Species evolution and adaptation over time allow organisms to respond to environmental changes, influencing biodiversity and ecosystem dynamics. For instance, the evolution of new plant species can affect soil composition, influencing resource availability for other organisms. Another significant biological process is ecosystem succession, where one community of organisms gradually replaces another over time. This natural progression transforms habitats, altering ecological relationships and the functioning of ecosystems. These processes are essential for maintaining biodiversity and the resilience of ecosystems.

3.1.2. Human-Induced Drivers of Environmental Transformation

Human activities have become dominant in driving environmental transformation, often resulting in far-reaching and unintended consequences. These impacts span various sectors, shaping ecosystems and influencing global environmental change (Emadodin and Bork, 2012; Essl et al., 2019; Mead et al., 2013).

One major area of influence is industrialization and urbanization. Resource extraction through industrial activities such as mining, logging, and fossil fuel extraction depletes natural resources, destroys habitats, and leads to pollution and biodiversity loss. Urban expansion, driven by the growth of cities and infrastructure, alters land use patterns, increases pollution, and significantly impacts local climates. Urban areas often experience heat island effects due to reduced green spaces and the concentration of infrastructure, further contributing to environmental stress.

Agricultural practices also have profound environmental impacts. Intensive agriculture, characterized by monoculture and chemical fertilizers and pesticides, leads to soil degradation, water pollution, and a reduction in biodiversity. Deforestation, often undertaken to make way for agricultural expansion, reduces wildlife habitats, alters local and global climate patterns, and disrupts critical carbon and water cycles.

Climate change is another significant consequence of human activities, particularly through the emission of greenhouse gases. The burning of fossil fuels, deforestation, and various industrial processes release large amounts of greenhouse gases into the atmosphere, contributing to global warming. As global temperatures rise, more frequent and severe weather events such as heatwaves, storms, and droughts occur, affecting both ecosystems and human societies.

Pollution from human activities also poses severe environmental challenges. Air pollution from vehicles, industries, and other sources not only harms human health but also damages ecosystems and exacerbates climate change. Water pollution, caused by agricultural runoff, industrial discharges, and sewage, threatens aquatic life and contaminates drinking water. Soil contamination, resulting from chemical use and improper waste disposal, reduces agricultural productivity and degrades ecosystem health.

Technological and economic factors play a dual role in environmental transformation. Technological advancements can positively impact the environment, as seen in renewable energy technologies that reduce reliance on fossil fuels. However, new industrial processes can increase resource consumption and waste. Economic growth and consumption patterns often lead to heightened demand for goods and services, further straining natural resources and intensifying environmental degradation.

3.1.3. Socio-Political Drivers

Socio-political factors play a crucial role in shaping environmental policies, priorities, and practices, significantly influencing the direction and pace of environmental transformation (Herrfahrdt-Pähle et al., 2020; Görg et al., 2017; Shkaruba et al., 2021). Policy and governance are foundational in this process. Government policies and environmental regulations are key drivers of environmental change, as they establish standards for pollution control, resource management, and conservation. Such policies can mitigate environmental degradation and promote sustainable practices when effectively implemented. Additionally, international agreements, like the Paris Agreement on climate change, shape national policies and foster global cooperation on environmental issues. These agreements serve as frameworks for countries to align their efforts in addressing global environmental challenges.

Public awareness and advocacy also have a profound impact. Environmental movements, often initiated by grassroots organizations, are instrumental in raising awareness about environmental issues, pushing for policy changes, and encouraging sustainable practices. These movements mobilize public support and influence political agendas. Furthermore, education and advocacy efforts, including public awareness campaigns, are critical in informing people about environmental issues. They help drive behavioral changes and influence policy decisions by increasing public engagement with environmental causes.

Economic incentives further shape environmental transformation by influencing both production and consumption patterns. Market forces, including economic incentives like subsidies for renewable energy or carbon pricing mechanisms, are essential levers in driving environmental progress. These financial tools can encourage the adoption of cleaner technologies and more sustainable business practices. Additionally, corporate responsibility has gained increasing attention, with businesses recognizing the importance of sustainability in their operations. The emphasis on corporate social responsibility (CSR) pushes companies to reduce their environmental footprints, adopt sustainable practices, and contribute positively to environmental transformation.

These socio-political factors—policy and governance, public advocacy, and economic incentives—create a dynamic framework supporting the transition to a more sustainable and environmentally responsible future.

3.1.4. Cultural and Behavioral Drivers

Cultural values and individual behaviors are pivotal in environmental transformation by shaping societal attitudes and practices toward the natural world.

Cultural attitudes are key determinants in how societies interact with and value the environment. Environmental ethics, rooted in cultural and ethical perspectives, influence how societies perceive the relationship between humans and nature. For example, some cultures emphasize stewardship and harmony with the environment, while others may prioritize economic growth over environmental preservation. Additionally, lifestyle choices—such as consumption habits, dietary preferences, and waste management practices—are deeply embedded in cultural norms (Shrivastava et al., 2020; Fernandes and Guiomar, 2016; O'Brien and Sygna, 2013). These behaviors can either promote or hinder environmental sustainability, depending on how societies prioritize resource use and environmental care.

Behavioral change is essential for fostering sustainable environmental practices. Whether undertaken by individuals or communities, sustainable practices can significantly contribute to environmental transformation.

Actions such as reducing waste, conserving energy, and choosing sustainable products are important steps toward achieving sustainability. Moreover, governments and organizations often use behavioral interventions to encourage these practices. Policies and programs designed to promote behavioral shifts, such as incentives for recycling, energy efficiency programs, or public campaigns encouraging conservation, can lead to substantial environmental improvements.

Together, cultural values and individual behaviors shape the trajectory of environmental transformation by influencing the way societies engage with environmental challenges and opportunities. These cultural and behavioral dimensions are integral to driving collective action toward a more sustainable future.

Drivers of environmental transformation encompass a wide range of factors, including natural processes, human activities, socio-political influences, and cultural attitudes. Understanding these drivers is essential for developing effective strategies to address environmental challenges and promote sustainable development. By analyzing the complex interactions between these factors, policymakers, businesses, and individuals can better navigate the path toward environmental resilience and sustainability.

3.2. Climate Change and Global Environmental Crises

Climate change and global environmental crises are deeply interconnected phenomena that pose significant challenges to ecosystems, human societies, and global sustainability. Understanding their causes, impacts, and potential solutions requires a comprehensive analysis of how they interact with various environmental and socioeconomic factors (von Storch et al., 2008; Singh, 2021; Singh and Singh, 2017).

3.2.1. Climate Change: Overview and Impacts

Climate change refers to the long-term shifts in temperature, precipitation, and other atmospheric conditions. The primary driver of current climate change is the rise in greenhouse gas emissions due to human activities, such as the burning of fossil fuels, deforestation, and various industrial processes.

3.2.1.1. Causes of Climate Change

The increase in greenhouse gas emissions is a major factor in climate change. The burning of fossil fuels like coal, oil, and natural gas releases large amounts of carbon dioxide (CO_2), methane (CH_4), nitrous oxide (N_2O), and other greenhouse gases into the atmosphere. These gases trap heat, causing the greenhouse effect and warming the Earth's surface. Deforestation is another contributing factor, as clearing forests for agriculture, logging, and urban development reduces the number of trees that naturally absorb CO_2. This further elevates atmospheric CO_2 levels (Mimura, 2013; Fraga et al., 2012; Jaramillo and Mendoza-Ponce, 2022). Additionally, industrial activities contribute to climate change through the combustion of fossil fuels and the release of gases used in industrial processes. Some of these activities also produce pollutants that affect the atmospheric balance, exacerbating climate-related issues.

3.2.1.2. Impacts of Climate Change

Climate change has increased global temperatures, resulting in more frequent and intense heatwaves. This warming poses serious threats to ecosystems, agriculture, and human health. It has also caused melting ice in glaciers and polar ice caps, which leads to rising sea levels. As sea levels rise, coastal communities face heightened flooding risks, erosion, and saltwater intrusion into freshwater resources. Changing weather patterns is another significant impact, as climate change drives more extreme and unpredictable weather events, such as hurricanes, droughts, and heavy rainfall. These changes disrupt agriculture, water supplies, and infrastructure, making it more difficult for societies to adapt. The increase in ocean acidification due to higher CO_2 absorption by oceans has a detrimental effect on marine life, particularly on organisms like shellfish and coral reefs, which are crucial for maintaining marine biodiversity and supporting fisheries. Furthermore, ecosystem disruption is a major consequence of climate change, as shifts in temperature and precipitation patterns affect species distributions, leading to biodiversity loss and altered ecological processes.

The causes and impacts of climate change are complex and interconnected, with far-reaching consequences for natural ecosystems, human societies, and the global environment. Addressing these challenges requires comprehensive action on both mitigation and adaptation.

3.2.2. Global Environmental Crises

Global environmental crises are interconnected issues that threaten the planet's health and human societies. These crises are often linked to climate change and can intensify its effects (Cooper et al., 2021; Tienhaara, 2010; Hamilton et al., 2015).

3.2.2.1. Biodiversity Loss

The loss of biodiversity is a major crisis driven by habitat destruction due to human activities like deforestation, urbanization, and agriculture. These actions lead to habitat fragmentation and push species toward extinction, reducing global biodiversity. Overexploitation of resources, such as overfishing, hunting, and resource extraction, further depletes wildlife populations and disrupts ecosystems, leading to declines in species numbers and overall ecosystem degradation. Introducing invasive species into non-native environments can outcompete indigenous species, disrupt ecosystems, and contribute to biodiversity loss.

3.2.2.2. Water Scarcity

Depletion of freshwater resources is a growing concern, as over-extraction for agricultural, industrial, and domestic uses depletes aquifers, rivers, and lakes, creating water scarcity and harming ecosystems. Pollution from agricultural runoff, industrial discharges, and waste further contaminates water sources, degrading water quality, harming aquatic life, and reducing the availability of safe drinking water. Climate change worsens these issues by altering precipitation patterns and increasing evaporation, exacerbating water shortages and threatening the water supply for communities and agriculture.

3.2.2.3. Land Degradation

Soil erosion caused by agricultural practices, deforestation, and construction reduces soil fertility, negatively affecting agricultural productivity. In arid and semi-arid regions, desertification becomes a critical issue, where once fertile land becomes increasingly barren and unproductive, harming ecosystems and the livelihoods of those who depend on the land. Urbanization contributes to land degradation by reducing arable land and natural habitats, disrupting local ecosystems, and altering local climates.

3.2.2.4. Waste Management

Improper waste management is a significant environmental crisis, particularly in the case of plastic pollution. The accumulation of plastic waste in oceans harms marine life, while microplastics enter the food chain, affecting both wildlife and human health. Electronic waste is another growing problem, as the disposal of electronic devices often leads to hazardous materials leaching into the environment, contaminating soil and water. Efficient waste management practices contribute to pollution, overflowing landfills, and greenhouse gas emissions. Efforts such as recycling and waste reduction are critical to sustainably managing waste.

Addressing these crises requires coordinated global efforts that focus on sustainable practices, conservation, and the protection of ecosystems to mitigate their effects and preserve the planet for future generations.

3.2.3. Interconnections and Feedback Loops

The interactions between climate change and global environmental crises create complex feedback loops that can amplify their impacts. For instance, climate change can exacerbate biodiversity loss by altering habitats and species distributions. Conversely, biodiversity loss can reduce ecosystem resilience, making it harder for ecosystems to adapt to climate change. Additionally, changes in precipitation and increased evaporation due to climate change can worsen water scarcity, affecting agricultural productivity and increasing competition for water resources. Deforestation and land degradation contribute to climate change by releasing stored carbon. Climate change can accelerate land degradation through increased temperatures and altered precipitation patterns. Furthermore, climate change impacts waste management through increased temperatures and extreme weather events that affect waste processing and disposal. Waste, mainly plastic pollution, can also impact the climate through greenhouse gas emissions from decomposition (Aït-El-Hadjait, 2017; Willems, 1997; Jolly et al., 2018).

3.2.4. Strategies for Addressing Climate Change and Environmental Crises

Addressing climate change and global environmental crises requires a multifaceted approach that includes mitigation, adaptation, and sustainable development strategies. Mitigation efforts focus on reducing greenhouse gas emissions to slow down the rate of climate change. This can be achieved by transitioning to renewable energy sources, improving energy efficiency, and implementing carbon pricing mechanisms. Additionally, conserving existing forests, restoring degraded lands, and expanding reforestation efforts can sequester carbon and support biodiversity. Sustainable agriculture practices, such as agroforestry, conservation tillage, and organic farming, can also reduce emissions and improve soil health (Munasinghe, 2011; Singh, 2021; Khojasteh et al., 2022).

Adaptation strategies aim to build resilience to the impacts of climate change. This involves implementing measures to enhance resilience to climate impacts, such as constructing flood defenses, developing drought-resistant crops, and improving water management. Strengthening disaster preparedness and response systems is crucial for managing extreme weather events and reducing their impacts on communities and infrastructure. Ecosystem-based adaptation, which utilizes natural systems to support adaptation efforts, can also be effective. Restoring wetlands for flood protection and preserving mangroves to reduce coastal erosion are examples of ecosystem-based adaptation.

Sustainable development strategies promote economic growth while minimizing environmental impact. A circular economy approach, which emphasizes reducing waste, reusing materials, and recycling, can significantly reduce environmental impact. Investing in and adopting green technologies, such as electric vehicles, energy-efficient appliances, and sustainable building materials, can also contribute to a more sustainable future. Global cooperation is essential for addressing global environmental issues collaboratively and sharing knowledge and resources. International agreements and partnerships can play a vital role in driving collective action and achieving shared goals.

Climate change and global environmental crises are interconnected challenges requiring comprehensive and integrated approaches. Understanding the drivers, impacts, and feedback loops associated with these issues is essential for developing effective mitigation, adaptation, and sustainable development strategies. By taking proactive measures and fostering global cooperation, it is possible to address these challenges and work towards a more resilient and sustainable future.

3.3. Technological Innovation and Green Transitions

Technological innovation is crucial in facilitating green transitions, shifting towards more sustainable and environmentally friendly practices across various sectors. Green transitions aim to reduce environmental impacts, enhance resource efficiency, and promote sustainable development. Technological advancements provide the tools and solutions necessary to achieve these goals, driving significant changes in how we produce, consume, and manage resources (Allam and Cheshmehzangi, 2024; Lin and Bai, 2023; Cooke, 2013).

3.3.1. Key Areas of Technological Innovation in Green Transitions

The transition to renewable energy sources is driven by technological advancements that improve efficiency, reduce costs, and expand the range of applications. Solar power has witnessed significant progress in renewable energy technologies, with advancements in photovoltaic (PV) technology leading to more efficient and affordable solar panels. Innovations such as flexible solar panels, solar roof tiles, and concentrated solar power (CSP) systems have further expanded the potential of solar energy. Similarly, wind energy has benefited from technological improvements in turbine design, resulting in larger, more efficient turbines that increase capacity and reliability. Offshore wind farms and floating turbines also expand the horizons for wind energy generation (Söderholm, 2020; Xu et al., 2023; Allam and Cheshmehzangi, 2024).

Hydropower, too, has seen innovations, with micro-hydropower systems providing energy solutions for small-scale and off-grid applications. Additionally, improvements in turbine design have enhanced efficiency and reduced environmental impacts. Geothermal energy has also witnessed advancements, with enhanced geothermal systems (EGS) improving the efficiency and feasibility of geothermal energy extraction, even in regions with lower geothermal potential.

Beyond renewable energy generation, energy storage and management are crucial for integrating renewable energy sources into the grid. Battery technology, particularly lithium-ion batteries and emerging technologies like solid-state and flow batteries, enhance energy storage capacity and efficiency. These advancements provide reliable storage solutions that support the integration of renewable energy sources. Smart grids, utilizing digital communication and control systems, optimize energy distribution, improve grid reliability, and facilitate the integration of renewable energy sources. These grids enable real-time monitoring and management of energy use. Advanced energy management systems (EMS), leveraging data analytics and artificial intelligence, optimize energy consumption, reduce waste, and improve efficiency in buildings, industrial processes, and transportation.

The transition to sustainable transportation is being driven by technological advancements that are reducing emissions, improving efficiency, and promoting cleaner transportation options. In the realm of electric vehicles (EVs), innovations in battery technology, charging infrastructure, and autonomous driving capabilities are expanding the appeal and practicality of EVs. Additionally, advancements in public transportation, such as electric buses, high-speed rail systems, and smart ticketing solutions, are enhancing the efficiency and sustainability of urban transit networks.

Alternative fuels, like hydrogen fuel cells and biofuels, also provide low-emission alternatives to conventional fossil fuels, supporting the transition to cleaner transportation options. In the context of the circular economy and waste management, technological advancements in recycling are improving the efficiency and effectiveness of recycling processes. Innovations in sorting systems and chemical recycling methods enhance the recovery of valuable materials from waste, supporting the circular economy.

Waste-to-energy technologies, which convert waste into energy through processes like anaerobic digestion and waste incineration with energy recovery, are helping to manage waste and generate renewable energy. Finally, eco-design, which involves creating products and systems with minimal environmental impact throughout their lifecycle, is promoted through materials science innovations and product design. These innovations encourage using sustainable materials and facilitate recycling and reuse, contributing to a more sustainable future.

3.3.2. Challenges and Opportunities

The widespread adoption of green technologies is often hindered by several barriers. One significant barrier is the initial cost associated with these technologies. While costs tend to decrease as technology matures and economies of scale are realized, the upfront investment can deter many. Additionally, the development of supporting infrastructure, such as charging stations for electric vehicles and grid integration for renewable energy, is crucial for successfully deploying green technologies. However, inadequate infrastructure can pose challenges (Bonaglia et al., 2024; Chen et al. 2023; Gliedt et al., 2018).

Furthermore, inconsistent or insufficient regulatory and policy frameworks can hinder the adoption of green technologies. Clear and supportive policies are necessary to promote innovation, investment, and widespread adoption of sustainable solutions. Integrating new technologies into existing systems and infrastructure can also be complex, requiring careful planning and coordination. Ensuring compatibility and addressing technical challenges are critical for successful implementation.

Despite these barriers, there are significant opportunities for advancement in the field of green technologies. Continued investment in research and development (RandD) is crucial for driving innovation and pushing the boundaries of technological capabilities. Collaborative efforts between governments, industry, and academia can accelerate technological progress and address key challenges. Public-private partnerships can also play a vital role in facilitating the deployment of green technologies by leveraging resources, expertise, and funding.

Global collaboration is essential for promoting, developing, and adopting green technologies across borders. International cooperation and knowledge sharing can help align efforts, address common challenges, and accelerate progress. Additionally, growing consumer demand for sustainable products and services can drive market growth and encourage companies to invest in green technologies. Raising awareness and educating consumers about environmental impacts can influence their choices and behavior, further driving the adoption of sustainable solutions.

3.3.3. Case Studies and Examples

Renewable energy has witnessed significant advancements, with solar power leading the way. Tesla's solar roof, which seamlessly integrates solar panels into roofing materials, offers an aesthetically pleasing and efficient residential solar energy generation solution. SunPower's high-efficiency solar panels also maximize energy generation in limited space, making solar energy more accessible for urban and residential applications (Truffer and Coenen, 2012).

Electric vehicles (EVs) have also gained traction, with Tesla's Model S pivotal in demonstrating the viability of EVs in the consumer market. The Model S's long-range capabilities, advanced battery technology, and performance have set a benchmark for the industry. In public transportation, BYD, a leading manufacturer of electric buses, has deployed its electric buses in various cities worldwide, showcasing the potential for sustainable transportation solutions.

The circular economy, which aims to minimize waste and maximize resource utilization, has also gained momentum. Patagonia, an outdoor apparel company, has embraced this approach by using recycled materials and encouraging customers to repair and recycle their clothing. TerraCycle, a company specializing in recycling solutions, offers innovative programs for hard-to-recycle materials, contributing to the circular economy by transforming waste into new products. These examples highlight the growing adoption of sustainable practices and the potential for technological advancements to drive a more circular and sustainable future (Söderholm, 2020).

Technological innovation is central to achieving green transitions and addressing environmental challenges. Advances in renewable energy, energy storage, sustainable transportation, and circular economy practices offer significant opportunities for reducing environmental impacts and promoting sustainability. However, overcoming barriers to adoption and leveraging opportunities for advancement are essential for maximizing the benefits of these technologies (Chen et al., 2023). By fostering collaboration, investing in research and development, and supporting supportive policies, society can accelerate the transition to a more sustainable and resilient future.

3.4. Socio-Economic Inequality and Environmental Justice

Socio-economic inequality and environmental justice are deeply intertwined issues that highlight the disparities in how ecological burdens and benefits are distributed among different social and economic groups. Understanding this relationship is crucial for addressing both social inequities and environmental sustainability (Carvalho et al., 2022; Bhandari and Hanna, 2022, 2022a; Bhandari, 2022).

3.4.1. Understanding Socio-Economic Inequality

Socio-economic inequality refers to disparities in income, wealth, education, and overall quality of life between different social groups. These disparities can result from economic policies, historical legacies, and systemic biases, leading to unequal access to resources and opportunities (Afshar et al., 2021; Fan et al., 2023; Betts-Davies et al., 2024). Socio-economic inequality encompasses various disparities affecting individuals' opportunities and life outcomes. Income inequality refers to variations in income levels among individuals or households. High-income individuals often have better access to resources and opportunities than low-income individuals. Wealth inequality, on the other hand, focuses on differences in asset ownership, such as property, investments, and savings. This type of inequality can perpetuate economic disparities across generations.

Educational inequality arises from disparities in access to quality education and educational resources. These disparities can affect individuals' ability to achieve economic mobility and gainful employment. Health inequality encompasses variations in health outcomes and access to healthcare services. Socioeconomic status can significantly influence individuals' health, including preventive care and treatment access. Lastly, employment inequality refers to differences in job opportunities, wages, and working conditions. Inequality in employment can affect economic stability and social mobility.

3.4.2. Understanding Environmental Justice

Environmental justice is a concept that addresses the fair distribution of environmental benefits and burdens. It ensures that all communities, especially marginalized and low-income groups, have equal protection from environmental hazards and equal access to environmental benefits.

The principles of environmental justice guide its implementation. Fair distribution ensures no group, particularly marginalized or low-income communities, bears disproportionate environmental harm or pollution. Meaningful participation provides opportunities for all communities to be involved in decision-making processes related to environmental policies and practices. Access to information ensures transparency and the availability of information about environmental risks, hazards, and policies affecting communities (Mohai et al., 2009; Figueroa, 2022; Sze and London, 2008). Protecting vulnerable populations recognizes and addresses the unique vulnerabilities of certain groups, including children, elderly individuals, and marginalized groups, as well as environmental risks.

3.4.3. Intersections of Socio-Economic Inequality and Environmental Justice

Environmental justice is a concept that addresses the unequal distribution of environmental benefits and burdens. Marginalized and low-income communities often bear a disproportionate share of environmental burdens, including exposure to pollution and vulnerability to climate change. These communities frequently face higher levels of exposure to environmental pollutants, such as air and water pollution, due to the proximity of industrial facilities, waste disposal sites, and high-traffic areas. Increased exposure to pollutants can lead to adverse health outcomes, including respiratory illnesses, cardiovascular diseases, and other health issues. Socio-economically disadvantaged groups may have limited access to healthcare services to address these impacts (Ryder, 2017; Malin et al., 2019; Alvarez and Evans, 2021).

Marginalized communities are often more vulnerable to the effects of climate change, including extreme weather events, sea-level rise, and temperature fluctuations. These communities may have fewer resources to adapt to or recover from climate-related impacts. Regarding environmental benefits, socio-economically disadvantaged neighborhoods may have limited access to green spaces, parks, and recreational areas. Lack of access to green spaces can impact physical and mental well-being. Additionally, lower-income communities may live in substandard housing with inadequate insulation, ventilation, or access to clean water and sanitation. Poor housing conditions can exacerbate environmental health risks. Access to sustainable resources, such as clean energy, healthy food, and efficient transportation, may also be limited for socio-economically disadvantaged groups. This lack of access can perpetuate inequalities and hinder efforts to promote sustainability.

3.4.4. Addressing Socio-Economic Inequality and Promoting Environmental Justice

Addressing environmental justice requires a multifaceted approach that involves policy and regulation, community engagement, data and research, and corporate responsibility. On the policy and regulatory front, implementing and enforcing environmental regulations is crucial to reduce pollution and ensure that environmental hazards do not disproportionately affect marginalized communities. Policies should prioritize equitable enforcement and mitigation. Additionally, providing economic incentives for sustainable practices, such as subsidies for renewable energy installations or green infrastructure, can help ensure that these benefits are accessible to low-income communities. Equitable development planning, which incorporates environmental justice principles into urban planning and development, is essential to prevent new projects and infrastructure from disproportionately burdening disadvantaged communities (Ryder, 2017; Malin et al., 2019; Alvarez and Evans, 2021).

Community engagement is another key component of addressing environmental justice. Participatory decision-making empowers marginalized communities to have a say in environmental decision-making processes, ensuring their voices and concerns are heard and addressed. Supporting community organizations and advocacy groups that work to address environmental justice issues and empower marginalized communities is also crucial. Raising awareness about environmental justice issues and educating communities about their rights and available resources through public education campaigns can help build support for equitable environmental policies.

Data and research are vital in identifying and addressing environmental justice concerns. Using geographic information systems (GIS) and data analytics, environmental justice mapping can help identify and map pollution sources, health outcomes, and socio-economic disparities. Research and analysis are also essential to understand the relationships between socioeconomic factors and environmental outcomes, informing policies and interventions to reduce disparities.

Corporate responsibility is another important aspect of addressing environmental justice. It is crucial to encourage businesses to adopt sustainable practices that minimize environmental impacts and promote equity in their operations and supply chains. Integrating environmental justice considerations into corporate social responsibility (CSR) initiatives, including support for community-based projects and investments in underserved areas, can further address environmental justice concerns (Arauco et al., 2014; Sidabalok, 2023; Sen and Azonobi, 2004).

Socio-economic inequality and environmental justice are interconnected issues that require comprehensive and equitable solutions. Addressing these challenges involves recognizing the disproportionate impacts of environmental burdens on marginalized communities, promoting fair access to environmental benefits, and ensuring meaningful participation in decision-making processes (Ryder, 2017; Malin et al., 2019; Alvarez and Evans, 2021). By implementing supportive policies, engaging communities, conducting research, and fostering corporate responsibility, society can work towards a more just and sustainable future for all.

3.5. Policy Shifts and Global Environmental Governance

Policy shifts and global environmental governance are essential in addressing complex environmental challenges and achieving sustainable development. These areas involve changes in policy frameworks and international cooperation to enhance environmental protection, mitigate climate change, and promote sustainable resource management (Biermann and Pattberg, 2008; Bernstein, 2004; Young, 2008).

3.5.1. Policy Shifts in Environmental Management

The evolving landscape of environmental policies reflects a shift towards proactive and integrated approaches. Traditional policies often focus on reactive measures to address environmental degradation. Contemporary policies emphasize prevention, early detection, and sustainability integration. These policies incorporate principles of sustainable development, aiming to balance economic growth, social equity, and environmental protection. This includes adopting practices that reduce ecological footprints and enhance resource efficiency. Additionally, stricter regulations are being implemented to address pollution, resource depletion, and biodiversity loss. These policies focus on enforcing compliance, setting higher standards, and promoting innovation in environmental protection technologies.

In the context of climate change, the Paris Agreement (2015) represents a significant policy shift towards global climate action. It establishes targets for reducing greenhouse gas emissions, promotes climate resilience, and emphasizes the need for financial and technological support to developing countries. Carbon pricing mechanisms, including carbon taxes and cap-and-trade systems, have also gained prominence as policy tools to internalize the environmental costs of carbon emissions. These mechanisms incentivize reductions in greenhouse gas emissions and support the transition to a low-carbon economy. Furthermore, policies promoting renewable energy sources, such as subsidies, tax credits, and feed-in tariffs, encourage developing and deploying clean energy technologies (Benson et al., 2013; Driessen et al., 2012; Lent and Wells, 1992). These incentives help reduce reliance on fossil fuels and support the growth of the renewable energy sector.

Biodiversity conservation is another critical area of focus in environmental policies. Expanding protected areas and strengthening conservation laws are key measures for preserving biodiversity. This includes creating national parks, wildlife reserves, and marine protected areas to safeguard habitats and species. Sustainable land use planning, which integrates biodiversity considerations into land use planning, helps balance development with conservation objectives. Policies promote sustainable agriculture, forest management, and urban planning that support biodiversity conservation. International biodiversity agreements, like the Convention on Biological Diversity (CBD), guide global efforts to conserve biodiversity, promote sustainable use of natural resources, and ensure fair and equitable sharing of benefits from genetic resources.

3.5.2. Global Environmental Governance

Global Environmental Governance is a framework for international cooperation and regulation addressing global environmental challenges. It involves developing, implementing, and enforcing international agreements, treaties, and standards to protect the planet's natural resources and ecosystems (Speth and Haas, 2007; Biermann and Pattberg, 2008; O'Neill et al., 2013).

Key aspects of global environmental governance include:

- *International Environmental Agreements*: These agreements, like the Paris Agreement on climate change and the Montreal Protocol, provide a platform for countries to collaborate on environmental issues.
- *International Organizations*: Organizations such as the United Nations Environment Programme (UNEP) and the Intergovernmental Panel on Climate Change (IPCC) are crucial in coordinating international efforts, conducting research, and offering policy advice.
- *Global Environmental Standards*: International standards, like those developed by the International Organization for Standardization (ISO), provide guidelines for industries and organizations to adopt sustainable practices and reduce their environmental impact.
- *Environmental Diplomacy*: Negotiations and cooperation between countries are essential for addressing transboundary environmental issues. Environmental diplomacy involves building consensus, resolving conflicts, and promoting sustainable development.
- *Public-Private Partnerships*: Collaboration between governments, businesses, and civil society organizations is crucial for effective global environmental governance. Partnerships can leverage resources, expertise, and innovation to address environmental challenges.

3.5.2.1. International Institutions and Agreements

- *United Nations Framework Convention on Climate Change (UNFCCC):* The UNFCCC plays a pivotal role in international climate negotiations and cooperation. Its annual Conferences of the Parties (COP) provide a platform for countries to negotiate and implement climate action plans, underscoring the importance of your participation in these global efforts.
- *Intergovernmental Panel on Climate Change (IPCC):* The IPCC assesses scientific knowledge on climate change and provides policy-relevant reports to guide global climate action. It plays a crucial role in informing international climate negotiations and policy development.

- *Global Environment Facility (GEF):* The GEF provides financial support for environmental projects that address global challenges, including climate change, biodiversity loss, and land degradation. It works with international organizations and national governments to implement sustainable development projects.

3.5.2.2. Multilateral Environmental Agreements (MEAs)

- *Convention on International Trade in Endangered Species (CITES):* CITES regulates international trade in endangered species to ensure that trade does not threaten their survival. It establishes trade controls and promotes conservation efforts for species at risk.
- *Basel Convention:* The Basel Convention addresses the transboundary movement and disposal of hazardous waste. It aims to minimize the generation of hazardous waste and promote environmentally sound management practices.
- *Stockholm Convention*: The Stockholm Convention focuses on eliminating or reducing persistent organic pollutants (POPs) that pose risks to human health and the environment. It sets guidelines for the management and disposal of these toxic substances.

3.5.2.3. Regional and Local Governance

- *Regional Agreements*: Regional environmental agreements address specific ecological challenges within geographic regions. Examples include the European Union's environmental policies, the African Union's environmental initiatives, and regional agreements on transboundary water management.
- *Local Governance:* Local governments play a critical role in implementing environmental policies and initiatives. They manage urban planning, waste management, and local conservation efforts, contributing to broader environmental goals.

3.5.2.4. Stakeholder Engagement and Collaboration

- *Public Participation*: Engaging stakeholders, including communities, businesses, and non-governmental organizations (NGOs), in environmental decision-making processes enhances transparency and accountability. Public participation helps ensure that policies address diverse needs and concerns.

- *Public-Private Partnerships*: Collaborations between governments and private sector entities can drive innovation and implementation of sustainable practices. Partnerships can focus on renewable energy, green infrastructure, and sustainable supply chains.
- *Scientific and Technical Expertise:* Leveraging scientific and technical expertise is crucial for effective environmental governance. Institutions, researchers, and experts provide valuable insights and data for policy development and implementation.

3.5.3. Challenges and Opportunities

Coordinating policies and actions across different levels of government and between countries can be challenging. Ensuring policy coherence and integration is essential for effectively addressing complex environmental issues. Additionally, ensuring that environmental policies and governance structures are equitable and inclusive is crucial. Addressing the needs of marginalized communities and ensuring fair distribution of benefits and burdens is critical for achieving environmental justice. Securing adequate funding and resources for environmental initiatives is often a challenge. Financial constraints can limit the implementation and effectiveness of policies and projects. Finally, ensuring compliance with environmental regulations and agreements is essential for achieving desired outcomes (Bernstein, 2004; Roberts, 2008; Li, 2006). Effective enforcement mechanisms are needed to address non-compliance and promote accountability.

3.5.3.1. Opportunities for Global Environmental Governance

Technological advancements offer opportunities to improve governance and address environmental challenges. Data monitoring, modeling, and communication innovations can enhance policy effectiveness and decision-making. Global environmental challenges require international cooperation and collaboration. Strengthening international agreements and partnerships can enhance collective efforts to address climate change and biodiversity loss. Integrating environmental considerations into broader policy frameworks, such as economic, trade, and development policies, can promote sustainable development and ensure that environmental goals are addressed across sectors.

Empowering communities and raising awareness about environmental issues can drive positive change and support the implementation of policies. Education and capacity-building initiatives can enhance public understanding and engagement in environmental governance.

Policy shifts and global environmental governance are critical in addressing environmental challenges and promoting sustainability. Evolving policies, international agreements, and effective governance structures are essential for managing environmental impacts, mitigating climate change, and achieving sustainable development goals. Governments, organizations, and stakeholders can work together to create a more sustainable and equitable future for all by addressing challenges and leveraging opportunities.

3.6. The Role of International Organizations and Agreements (UNFCCC, IPBES, etc.)

International organizations and agreements are crucial in shaping global environmental policies, facilitating international cooperation, and addressing transboundary environmental issues. They provide collective action, scientific assessment, and policy development frameworks to address pressing environmental challenges (DeSombre, 2010; Stephan andZelli, 2018; Breitmeier, 1997). Here is a detailed look at some key international organizations and agreements:

3.6.1. United Nations Framework Convention on Climate Change (UNFCCC)

This was established in 1992 to stabilize greenhouse gas concentrations in the atmosphere at a level that prevents dangerous interference with the climate system (Bodansky, 1993; Sands, 1992; Hickmann et al., 2021).

3.6.1.1. Key Functions

- *Climate Negotiations:* The UNFCCC hosts annual Conferences of the Parties (COP), where member countries negotiate and implement climate action plans. Major agreements include the Kyoto Protocol (1997) and the Paris Agreement (2015).

- *Reporting and Monitoring*: The UNFCCC requires parties to report on their greenhouse gas emissions, climate actions, and progress towards targets. This transparency helps track progress and ensures accountability.
- *Financial and Technical Support*: The UNFCCC provides financial mechanisms, such as the Green Climate Fund (GCF), to support climate action in developing countries. It also facilitates technology transfer and capacity-building efforts.

3.6.1.2. Impact

- *Global Climate Policy*: The UNFCCC framework has significantly influenced global climate policy, driving commitments to reduce emissions and enhance climate resilience.
- *Global Agreements*: The Paris Agreement, a landmark achievement under the UNFCCC, sets global targets for limiting temperature rise and encourages countries to submit Nationally Determined Contributions (NDCs).

3.6.2. Intergovernmental Panel on Climate Change (IPCC)

This was established in 1988 to provide policymakers with regular assessments of scientific knowledge on climate change, its impacts, and potential adaptation and mitigation strategies (Change, 1990; Bhandari, 2018, 2022).

3.6.2.1. Key Functions

- *Assessment Reports*: The IPCC publishes comprehensive assessment reports that synthesize the latest scientific research on climate change. These reports inform policy decisions and shape international climate agreements.
- *Special Reports*: The IPCC also produces special reports on specific topics, such as the impacts of global warming of 1.5°C on land use, the ocean, and the cryosphere.
- *Workshops and Conferences*: The IPCC organizes workshops and conferences to engage with scientists, policymakers, and stakeholders

and facilitate knowledge exchange and discussions on climate science.

3.6.2.2. Impact

- *Scientific Consensus*: The IPCC's assessments provide a consensus view of climate science, influencing international climate negotiations and public understanding of climate issues.
- *Policy Guidance*: The IPCC's reports offer policymakers actionable recommendations for addressing climate change, supporting evidence-based decision-making.

3.6.3. International Union for Conservation of Nature (IUCN)

This was established in 1948 to promote nature conservation and sustainable use of natural resources (Bhandari, 2012, 2022; Upadhyaya, 2019).

3.6.3.1. Key Functions

- *Red List of Threatened Species*: The IUCN maintains the Red List, a comprehensive database of the conservation status of species worldwide. This tool helps prioritize conservation efforts and assess species extinction risk.
- *Protected Areas*: The IUCN develops guidelines and standards for establishing and managing protected areas, contributing to global biodiversity conservation.
- *Policy Advocacy*: The IUCN advocates for policies that support conservation and sustainable development, working with governments, NGOs, and other stakeholders.

3.6.3.2. Impact

- *Conservation Efforts:* The IUCN's work has helped establish and manage protected areas and mobilize global support for biodiversity conservation.

- *Biodiversity Monitoring*: The Red List is a critical resource for monitoring trends in species populations and guiding conservation priorities.

3.6.4. Convention on Biological Diversity (CBD)

This was established in 1992 to conserve biodiversity, promote sustainable use of its components, and ensure the fair and equitable sharing of benefits arising from genetic resources (Bhandari, 2020; De Oliveira et al., 2011; Chandra and Idrisova, 2011).

3.6.4.1. Key Functions

- *Global Biodiversity Targets*: The CBD sets global targets for biodiversity conservation, such as the Aichi Biodiversity Targets, which aim to address key drivers of biodiversity loss and protect ecosystems.
- *National Biodiversity Strategies*: The CBD encourages countries to develop and implement national biodiversity strategies and action plans tailored to their needs and priorities.
- *Financial Mechanisms*: The CBD supports funding mechanisms, such as the Global Environment Facility (GEF), to assist countries in implementing biodiversity conservation projects.

3.6.4.2. Impact

- *Biodiversity Conservation*: The CBD has significantly advanced global biodiversity conservation efforts and promoted sustainable practices.
- *International Collaboration:* The CBD fosters international collaboration on biodiversity issues, facilitating the exchange of knowledge and resources among member countries.

3.6.5. The Basel Convention

This was established in 1989 to control and reduce the transboundary movement of hazardous waste and ensure its environmentally sound management (Kummer, 1998, 1999; Krueger, 2013).

3.6.5.1. Key Functions

- *Waste Management Guidelines*: The Basel Convention provides guidelines for the handling, treating, and disposing of hazardous waste, aiming to minimize environmental and health risks.
- *Technical Assistance*: The Convention offers technical assistance and capacity-building support to countries in managing hazardous waste and improving waste management systems.
- *International Cooperation*: The Basel Convention facilitates international cooperation to address the global challenges of hazardous waste management and reduce illegal waste trafficking.

3.6.5.2. Impact

- *Waste Management Standards*: The Basel Convention has contributed to establishing international standards for hazardous waste management and has helped reduce environmental contamination from waste.
- *Global Awareness*: The Convention raises awareness about the risks associated with hazardous waste and promotes practices that minimize its environmental impact.

3.6.6. The Stockholm Convention

This was established in 2001 with the objective of eliminating or restricting the production and use of persistent organic pollutants (POPs) that pose risks to human health and the environment (Lallas, 2001; Hagen and Walls, 2005; Tang, 2013).

3.6.6.1. Key Functions

- *POPs Elimination:* The Stockholm Convention sets guidelines for the elimination of POPs, including substances such as DDT and PCBs. It supports efforts to phase out these harmful chemicals and manage their legacy.
- *Technical Assistance:* The Convention provides technical assistance and capacity-building to help countries implement POPs management measures and develop alternatives.
- *Monitoring and Reporting*: The Convention promotes monitoring and reporting of POP levels in the environment, helping to track progress and assess the effectiveness of control measures.

3.6.6.2. Impact

- *Reduction of Toxic Chemicals*: The Stockholm Convention has led to significant reductions in the use and release of toxic POPs, contributing to improved environmental and human health.
- *International Cooperation*: The Convention fosters global cooperation in addressing the challenges posed by persistent pollutants and supports efforts to find safer alternatives.

International organizations and agreements play a pivotal role in global environmental governance by providing frameworks for cooperation, setting standards, and supporting implementation efforts. Their work is essential for addressing transboundary environmental issues, advancing sustainability goals, and promoting equitable and effective solutions to environmental challenges. By facilitating international collaboration and providing technical and financial support, these organizations help drive progress toward a more sustainable and resilient future for the planet.

Chapter 4

Key Stakeholders in the Transformation of Environmentalism

The transformation of environmentalism involves a diverse array of stakeholders, each playing a crucial role in shaping policies, driving change, and implementing strategies for sustainability. Understanding the roles and contributions of these stakeholders is essential for fostering effective environmental transformation (Poncelet, 2001; Saleem et al., 2020; Miller, 2008).

4.1. Governments and Policy Makers

Governments and policymakers are pivotal in environmentalism transformation due to their authority to create, implement, and enforce environmental policies and regulations. Their decisions influence national and global environmental strategies, and their leadership is crucial for addressing complex environmental challenges (Horan, 2019; Sachs et al., 2019; Henríquez, 2014). Here is a detailed look at their roles and contributions:

4.1.1. The Role of National Governments

Policy formulation and implementation play a critical role in addressing environmental challenges at the national level. Governments develop legislative frameworks, such as the Clean Air Act in the United States and the Environment Protection Act in India, which serve as the foundation for environmental protection. These laws address key issues like pollution control, wildlife preservation, and land use management. Regulations and standards are also established to guide industries in managing their environmental impact, including limits on emissions, waste management, and resource use. Regulatory frameworks typically include provisions for environmental impact assessments and set compliance requirements. Effective enforcement of these regulations is essential, with governments conducting monitoring, inspections, and imposing penalties to ensure compliance and deter violations (Aydın, 2016; Kettl, 2000).

Strategic planning and integrating environmental considerations into broader development policies are crucial for achieving sustainability. Governments create national environmental strategies and action plans that outline goals, priorities, and specific measures for environmental protection. These plans are often aligned with international environmental commitments while addressing local needs. Moreover, environmental considerations are integrated into national development policies related to economic growth, infrastructure projects, and social policies. This ensures that environmental sustainability is woven into the broader fabric of national development strategies, enhancing the long-term viability of growth initiatives.

Resource allocation and funding are vital components of effective environmental governance. National governments allocate budgets to support environmental programs, such as conservation efforts, climate adaptation projects, and pollution control measures. These funds are directed toward research, infrastructure development, and the implementation of environmental policies. Additionally, governments provide economic incentives, such as subsidies, tax credits, and grants, to encourage sustainable practices and the adoption of green technologies. For example, subsidies for renewable energy projects help promote investment in clean energy, supporting the transition to a low-carbon economy. Effective resource allocation ensures that governments have the financial tools necessary to support their environmental objectives.

4.1.2. The Role of Local Governments

Local environmental management is vital for addressing community-specific issues and ensuring sustainability at the grassroots level. Local governments are responsible for managing waste collection, recycling programs, and pollution control measures, focusing on localized solutions for problems like air and water quality. They also oversee land use planning and urban development, ensuring that projects comply with environmental standards and contribute to sustainable urban growth (Brand and Thomas, 2013; Hodson and Marvin, 2017; Kettl, 2000). By controlling zoning regulations and managing land use, local authorities play a crucial role in balancing development with environmental preservation.

Community engagement and education are key elements of local environmental management. Local governments organize public awareness campaigns to educate residents about environmental issues and promote sustainable practices such as recycling, energy conservation, and water management. These campaigns aim to foster community participation in environmental conservation efforts. Additionally, local authorities actively involve communities in addressing environmental concerns, seeking input on local policies, and encouraging residents to participate in decision-making processes. This inclusive approach ensures that environmental solutions are tailored to the specific needs and priorities of local populations.

The implementation of national environmental policies often falls to local governments, who adapt these policies to regional contexts. They are responsible for executing national environmental regulations and addressing specific local challenges. Local governments often establish additional regulations and initiatives to complement national efforts, targeting unique environmental issues that their communities face. This localized execution of national policies ensures that environmental governance is flexible and responsive to the diverse needs of different regions.

4.1.3. The Role of International Policy Makers

Global environmental agreements are crucial for addressing environmental challenges that cross national borders. International policymakers negotiate and implement agreements like the Paris Agreement on climate change and the Convention on Biological Diversity (CBD), establishing global standards and commitments to protect the environment. These agreements serve as frameworks for countries to work together on pressing environmental issues, setting collective goals and timelines for mitigating climate change, preserving biodiversity, and promoting sustainable development. This process establishes international standards, helping to unify global efforts toward environmental sustainability (Bäckstrand, 2003; Rootes, 1999; Jamison, 2001).

Coordination and cooperation are essential for the success of global environmental agreements. Policymakers facilitate collaboration between countries to address transboundary environmental issues, ensuring that collective action is taken to combat challenges like pollution, deforestation, and climate change. This collaboration often involves sharing resources, knowledge, and expertise, allowing countries to pool their efforts and work more effectively. By fostering international cooperation, global agreements enable countries to tackle environmental problems that no single nation can solve.

International support mechanisms are vital in helping countries meet their environmental commitments. Policymakers oversee financial mechanisms like the Global Environment Facility (GEF) and Climate Investment Funds (CIF), which provide funding for environmental projects, especially in developing countries. These funds help countries finance climate mitigation and adaptation efforts, biodiversity conservation, and pollution control initiatives. Additionally, international policymakers offer technical assistance and capacity-building to support countries in implementing environmental policies. This assistance includes access to technology, knowledge, and best practices, enabling countries to strengthen their environmental governance and achieve sustainability goals.

4.1.4. Challenges and Opportunities for Governments and Policy Makers

Governments face significant challenges in implementing environmental policies, often due to political and economic constraints. Balancing economic growth with environmental protection is particularly difficult, as it requires navigating competing interests and trade-offs between development and sustainability. Political pressures from various stakeholders, including businesses and industries, can sometimes slow down or weaken environmental initiatives. Additionally, achieving policy coherence is a persistent challenge. Environmental policies must align with other national priorities, such as economic development and energy production (Rootes, 1999; Chertow, 1997). However, these sectors often have conflicting goals, making it difficult to integrate environmental considerations into broader policy frameworks without compromising on either side.

On the other hand, there are notable opportunities that governments can seize to enhance environmental governance. Technological innovation offers immense potential for improving environmental management and promoting sustainable practices. Governments can advance their environmental goals while fostering economic growth by investing in green technologies—such as renewable energy, waste management systems, and sustainable agriculture. Innovation can also make industries more efficient and less harmful to the environment. Furthermore, public participation is a key opportunity for improving the effectiveness of environmental policies. Engaging citizens and stakeholders in the decision-making process helps build widespread support for environmental initiatives and ensures that policies are better tailored to the needs and priorities of local communities (Bäckstrand, 2003; Rootes, 1999; Jamison, 2001). Participatory approaches strengthen governance by making it more inclusive, transparent, and responsive to the public's environmental concerns.

Governments and policymakers are central to environmentalism transformation, as they hold the power to shape environmental policies, allocate resources, and implement regulations. Their role encompasses policy formulation, strategic planning, enforcement, and international cooperation. Addressing challenges and seizing opportunities can drive significant progress in environmental protection and sustainability, ultimately contributing to a more resilient and sustainable future.

4.2. Businesses and Industries

Businesses and industries play a crucial role in environmentalism transformation due to their significant impact on natural resources, emissions, and waste generation. Their practices, policies, and innovations can drive positive environmental change or contribute to environmental degradation (Shrivastava and Scott, 1992; Welford, 2013).

Corporate sustainability initiatives are increasingly becoming a core part of business operations, with companies adopting various strategies to minimize their environmental impact. One key approach is the implementation of sustainable business practices. Businesses focus on resource efficiency by optimizing energy use, reducing water consumption, and minimizing waste from raw materials. Examples include energy-efficient manufacturing processes and the establishment of zero-waste production systems that ensure minimal environmental harm. Another important aspect is the development of sustainable supply chains. Companies are committed to sourcing raw materials responsibly, reducing their environmental footprint, and ensuring ethical labor practices. Many businesses also seek certifications such as Fair Trade or incorporate recycled materials to enhance the sustainability of their supply chains.

Corporate Social Responsibility (CSR) is vital in promoting environmental stewardship within businesses. Many companies have CSR programs that focus on protecting the environment through initiatives like tree planting, wildlife conservation, and organizing community clean-up projects.

These efforts demonstrate a company's commitment to environmental protection beyond regulatory requirements. In addition to CSR programs, businesses are increasingly prioritizing transparency by publishing sustainability reports. These reports disclose their environmental impact, progress on sustainability goals, and compliance with environmental regulations, enhancing both accountability and public trust.

Innovation and the development of green technologies are essential for driving corporate sustainability. Many businesses are investing in renewable energy technologies, such as solar panels, wind turbines, and bioenergy systems, which help reduce reliance on fossil fuels and lower greenhouse gas emissions. Additionally, companies are producing and promoting energy-efficient products, such as LED lighting, high-efficiency appliances, and electric vehicles. These products not only reduce energy consumption but also mitigate environmental damage caused by traditional alternatives.

Research and development (RandD) are central to the advancement of sustainable solutions. Companies invest in creating innovative technologies and processes that address pressing environmental challenges, such as improving waste management systems, developing new materials, and enhancing recycling practices. Collaboration with academic institutions and research organizations is another critical aspect of RandD in sustainability. Through these partnerships, businesses can accelerate the development of green technologies and promote knowledge transfer, driving further innovation in corporate sustainability practices.

4.2.1. Environmental Impact Management

Environmental impact management is a critical aspect of corporate responsibility, requiring businesses to address their environmental footprint actively. One key area is compliance and regulation, where companies must adhere to environmental laws and standards governments and international bodies set. This includes meeting emission limits, managing hazardous waste, and conducting environmental impact assessments to ensure sustainable operations. Compliance is not only a legal obligation but also essential for maintaining a company's reputation and avoiding penalties. In addition, companies implement risk management strategies to identify potential environmental risks and mitigate their impact (Aronczyk, 2018; Hodson and Marvin, 2017; O'Riordan, 2013). This involves developing contingency plans for environmental incidents, such as spills or accidents, and adopting preventive measures to minimize the likelihood of such events.

Pollution reduction and waste management are equally important components of environmental impact management. To reduce pollution, businesses adopt technologies and practices that minimize emissions into the air and water. This may involve installing pollution control devices, optimizing industrial processes, or switching to cleaner production methods. Waste management is another area where companies can make a significant environmental impact. By implementing recycling programs, composting organic waste, and utilizing waste-to-energy systems, businesses can reduce the amount of waste they generate and manage it in an environmentally responsible way. These efforts reduce the strain on landfills and contribute to a more circular economy, where resources are reused and repurposed rather than discarded.

4.2.2. Industry Associations and Initiatives

Industry standards and certifications play a key role in promoting sustainability across various sectors. Many industries have developed voluntary environmental standards that encourage companies to improve their environmental performance. Certifications such as ISO 14001, which focuses on environmental management systems, and LEED (Leadership in Energy and Environmental Design), which promotes sustainable building practices, are widely recognized for setting benchmarks in environmental stewardship. These standards provide companies with frameworks to reduce their environmental impact and ensure compliance with best practices in sustainability. Additionally, industry associations often lead collaborative initiatives to tackle shared environmental challenges. These initiatives may focus on reducing carbon emissions, enhancing resource efficiency, and promoting sustainable practices throughout the industry. By working together, businesses can address broader environmental concerns and drive collective progress in sustainability (Park and Roome, 2017; Penna and Geels, 2012; Lozano, 2022).

Public-private partnerships are another critical avenue for addressing environmental issues. Businesses collaborate with governments to implement sustainability projects and initiatives aimed at environmental protection. These partnerships allow companies and governments to pool resources, expertise, and funding to achieve shared environmental goals, such as reducing pollution, promoting renewable energy, or improving waste management systems. Furthermore, engagement with non-governmental organizations (NGOs) is essential to many companies' sustainability strategies. By partnering with NGOs and environmental advocacy groups, businesses can support conservation efforts, develop more sustainable practices, and address social and environmental issues. These collaborations often enhance the credibility and impact of corporate sustainability initiatives, benefiting both the environment and society.

4.2.3. Challenges and Opportunities for Businesses and Industries

Businesses face several challenges in balancing profitability with environmental sustainability. Implementing sustainable practices often requires significant upfront investment, whether through adopting new technologies, redesigning supply chains, or modifying operational processes. These changes can lead to higher costs, making balancing economic objectives with sustainability goals difficult. For many companies, tiny and medium-sized enterprises, this challenge is compounded by the pressure to maintain profitability in competitive markets. Additionally, regulatory compliance presents its own set of difficulties. Environmental regulations are often complex and vary widely across different countries and regions. Multinational companies, in particular, must navigate a patchwork of regulatory environments, making ensuring compliance across their global operations challenging. Maintaining evolving regulations requires significant resources, including legal expertise and management systems (Shrivastava and Scott, 1992; Jansen and Vellema, 2004).

Despite these challenges, there are substantial opportunities for businesses that embrace sustainability. Companies adopting sustainable practices can gain a competitive advantage by appealing to environmentally conscious consumers, improving their brand reputation, and accessing new markets prioritizing sustainability. As more consumers and investors prioritize environmental responsibility, companies that lead in green innovation are well-positioned to differentiate themselves in the marketplace. Additionally, sustainable practices often lead to long-term cost savings. Investing in energy efficiency, waste reduction, and resource management can reduce operating costs over time by lowering resource consumption and improving operational efficiencies. As a result, businesses that commit to sustainability not only benefit the environment but also improve their financial performance in the long run.

Businesses and industries are key stakeholders in environmentalism transformation, with the potential to drive significant positive change through sustainable practices, innovation, and collaboration. By adopting green technologies, managing environmental impacts, and engaging in industry initiatives, they contribute to environmental protection and sustainability. Addressing challenges and seizing opportunities in this sector are crucial for advancing environmental goals and promoting a more sustainable future.

4.2.3. NGOs and Civil Society

Non-governmental organizations (NGOs) and civil society groups are vital in transforming environmentalism. They often act as catalysts for change, advocates for environmental protection, and mobilizers of public and political support. Their diverse activities encompass advocacy, research, community engagement, and direct action (Ho, 2001; Aydın, 2016; Finger, 2013; Bhandari, 2014; Bhandari and Oli, 2018).

Advocacy and awareness-raising are central to the work of environmental non-governmental organizations (NGOs). In policy advocacy, NGOs play a crucial role in influencing environmental legislation at local, national, and international levels. They lobby governments, engage in policy discussions, and advocate for stronger environmental regulations and standards. By actively participating in the policy-making process, NGOs help shape laws that protect the environment. Public campaigns are another critical tool NGOs use to raise awareness about pressing environmental issues like climate change, deforestation, and pollution. These campaigns aim to educate the public, inspire behavioral changes, and pressure policymakers to take action on environmental concerns.

In addition to advocacy, public education is a major focus of many environmental NGOs. They develop and deliver educational programs designed to inform the public about environmental issues and sustainable practices. These programs may take the form of workshops, seminars, or informational resources tailored to different audiences. NGOs also use media and communication platforms, including social media, websites, and publications, to spread information, share success stories, and engage a wider audience in environmental action. Through these efforts, they build public awareness and foster a more environmentally conscious society.

Research and expertise are another cornerstone of NGO efforts. Many environmental NGOs conduct or fund scientific research on topics such as biodiversity loss, the impacts of climate change, and pollution. This research generates valuable data and insights that can be used to support advocacy efforts and inform policy decisions. NGOs also conduct policy analysis to evaluate the effectiveness of current environmental regulations, identifying gaps and making evidence-based recommendations for improvement. In doing so, they provide critical insights into how policies can be enhanced to better address environmental challenges.

NGOs often act as expert consultants, offering technical advice and guidance on environmental matters. They collaborate with governments, businesses, and other stakeholders to share best practices and sustainable solutions. Additionally, many NGOs engage in participatory research, involving local communities and stakeholders in the research process (Ho, 2001; Aydın, 2016; Finger, 2013; Bhandari, 2014; Bhandari and Oli, 2018). This approach ensures that diverse perspectives are included in environmental initiatives, fostering more inclusive and effective solutions.

4.2.4. Community Engagement and Empowerment

Grassroots mobilization is a key strategy for many environmental NGOs as they implement community-based projects that address local environmental challenges. These initiatives can include local conservation efforts, clean-up campaigns, and tree-planting programs, all of which engage community members and encourage active participation in protecting their environment. By focusing on localized issues, NGOs can tailor their approaches to the specific needs and conditions of the communities they serve. Additionally, NGOs empower local leadership by training and supporting community leaders and activists. This helps build the capacity of communities to take ownership of environmental initiatives and sustain long-term advocacy and action at the grassroots level (Sawin and Zehr, 2013; Rudito et al., 2022).

Facilitating participation is another crucial role played by NGOs. They promote stakeholder engagement by fostering collaboration among different groups, including communities, governments, and businesses. By bringing diverse stakeholders together, NGOs ensure that environmental solutions are inclusive and reflect the needs of various groups. They also actively gather community feedback on environmental issues and policies, using this input to refine their initiatives and make sure they remain relevant and effective. This participatory approach not only improves the outcomes of environmental projects but also strengthens the bonds between NGOs and the communities they serve.

Direct action and conservation are central components of many NGOs' works. On-the-ground action, such as habitat restoration, wildlife protection, and combating illegal poaching, directly contributes to preserving ecosystems and species. These efforts are vital for maintaining biodiversity and ensuring the health of natural environments. In addition to conservation work, NGOs often address environmental justice issues by advocating for marginalized communities disproportionately affected by environmental degradation. This can involve fighting against pollution, defending land rights, and ensuring fair access to natural resources, thereby addressing the intersection of environmental and social justice.

NGOs also led advocacy campaigns to mobilize public support and drive policy changes on critical environmental issues. These campaigns often include organizing protests, circulating petitions, and building coalitions to amplify their impact. In many cases, NGOs participate in global environmental movements and collaborate with international organizations to address transboundary challenges like climate change, deforestation, and biodiversity loss (Sawin and Zehr, 2013; Rudito et al., 2022). NGOs play an influential role in shaping local and international environmental policies through their campaigns and global partnerships.

4.2.5. Challenges and Opportunities for NGOs and Civil Society

NGOs face significant challenges in their efforts to address environmental issues. One major hurdle is securing adequate funding and resources. Since many NGOs rely heavily on donations, grants, and external funding, their financial support can be unpredictable, which limits their ability to scale up initiatives or sustain long-term projects. This uncertainty often forces NGOs to operate with limited capacity, making it harder to meet growing environmental challenges (Pearce, 2000; Scholte, 2018). NGOs in some regions also encounter political and legal barriers that restrict their operations. Governments may impose restrictions on freedom of expression, limit advocacy work, or create bureaucratic hurdles that hinder NGOs' ability to advocate effectively and implement initiatives. These challenges can slow progress and reduce the impact of their work.

Despite these obstacles, NGOs also have significant opportunities to advance their environmental missions. Growing public awareness and concern about environmental issues, such as climate change, biodiversity loss, and pollution, provide NGOs with new avenues to engage broader audiences. As more people become conscious of environmental threats, NGOs can build stronger support networks, increase public participation, and drive greater collective action. Furthermore, collaboration and partnerships offer substantial opportunities for NGOs to amplify their impact. By working alongside governments, businesses, and other organizations, NGOs can pool resources, share expertise, and achieve common environmental goals more effectively. These partnerships allow NGOs to extend their reach and influence, creating a more coordinated and impactful approach to addressing global environmental challenges.

NGOs and civil society groups contribute to environmentalism transformation through advocacy, research, community engagement, and direct action. They are crucial in raising awareness, influencing policy, empowering communities, and implementing conservation initiatives. NGOs and civil society can drive meaningful progress in environmental protection and sustainability by addressing challenges and seizing opportunities, fostering a more inclusive and practical approach to addressing global environmental challenges.

4.2.6. Indigenous Communities and Local Actors

Indigenous communities and local actors are essential stakeholders in environmentalism transformation due to their deep-rooted connections with their environments, traditional knowledge, and community-based approaches to sustainability. Their involvement is crucial for achieving effective and equitable environmental solutions (Kasih et al., 2021; Lerner et al., 2017).

4.2.6.1. Indigenous Communities

Indigenous communities play a vital role in environmental stewardship, relying on traditional knowledge and practices passed down through generations to manage resources sustainably. Their agriculture, fishing, and forest management methods are often designed to maintain ecological balance and ensure that natural resources are used responsibly. These practices protect ecosystems and foster resilience in the face of environmental changes. Furthermore, indigenous cultures are deeply connected to their natural environments (Kasih et al., 2021; Lerner et al., 2017). This cultural and spiritual connection to nature shapes their conservation and environmental management approach, promoting a philosophy of respect and preservation that goes beyond mere resource use. Their knowledge systems offer valuable insights into sustainable living that are increasingly recognized for relevance in modern conservation efforts.

Indigenous communities strongly advocate for their land rights, critical for preserving their traditional territories and the ecosystems they sustain. Secure land rights enable these communities to manage their environments according to their cultural values and traditional practices. Indigenous groups also seek representation in environmental decision-making processes to ensure their perspectives and rights are considered (Reo et al., 2017; Steinveg, 2023). By participating in policy discussions, environmental assessments, and conservation planning, they aim to protect their lands and way of life while contributing to broader environmental goals.

Conservation and management are integral to indigenous communities' interaction with their environments. Many of these communities manage Indigenous Protected and Conserved Areas (IPCAs), significant for biodiversity conservation. These protected areas are governed using traditional practices emphasizing sustainability and long-term ecological health. Indigenous knowledge is crucial in preserving biodiversity by fostering diverse ecosystems and protecting species. Through their conservation efforts, Indigenous communities contribute to protecting their cultural heritage, global biodiversity, and environmental sustainability.

4.2.6.2. Local Communities

Local communities play a pivotal role in community-based environmental management, often implementing grassroots initiatives such as reforestation, water conservation, and waste management. These efforts are tailored to address the community's specific environmental challenges, ensuring that the solutions are both relevant and effective. Community members become directly involved in sustainable practices by engaging in these activities, fostering a sense of ownership and responsibility for local environmental stewardship. In addition, local knowledge—such as insights into climate patterns, soil conditions, and species behaviors—provides practical, context-specific solutions that more significant, more generalized programs may overlook (Miranda, 2010; Radcliffe, 2012).

Participatory approaches ensure that local actors are included in environmental decision-making. When local communities shape policies and initiatives, the results are more effective and better adapted to the local context. Their participation empowers them and ensures that environmental strategies reflect their unique needs and circumstances. Furthermore, feedback from local communities is crucial for continuously adapting and improving environmental policies and projects. Policymakers and organizations can gather valuable insights by engaging with communities, allowing for more responsive and targeted environmental management.

Sustainable livelihoods are often at the heart of community-based environmental efforts. Many local communities participate in eco-friendly economic activities, such as eco-tourism, sustainable agriculture, and community-based fisheries. These initiatives help balance the need for economic development with the protection of the environment, offering a pathway to long-term sustainability. Additionally, local communities often integrate cultural preservation with environmental management. They can address modern environmental challenges by maintaining traditional practices and knowledge while safeguarding their cultural heritage. This dual focus on economic sustainability and cultural preservation strengthens the community's resilience and ensures a holistic approach to environmental conservation.

4.2.6.3. Collaboration and Partnerships

Indigenous and local communities often engage in collaborative environmental projects through partnerships with NGOs, government agencies, and businesses. These collaborations unite diverse expertise, resources, and perspectives, creating a powerful synergy to address environmental issues. By working together, communities can leverage external organizations' technical knowledge and financial resources while ensuring that local knowledge and traditions are incorporated into project planning and execution. These joint initiatives frequently focus on community-led conservation efforts, the co-management of protected areas, and the design of sustainable development programs that reflect the needs and aspirations of all stakeholders involved (El Ansari et al., 2001; Schroeder, 1999; Carnwell and Carson, 2005). The collective effort enhances the success and sustainability of environmental projects.

In terms of policy influence, indigenous and local communities actively advocate for recognizing their rights and integrating their knowledge systems into broader environmental management frameworks. Through advocacy, these communities can shape policies that reflect their values and priorities, leading to more inclusive and effective environmental governance. Their participation in policy discussions allows them to directly influence environmental legislation, ensuring that it addresses their communities' ecological and cultural needs. By contributing their perspectives and experiences, these communities help create policies that are grounded in local realities and are more likely to succeed in achieving sustainable environmental outcomes.

4.2.6.4. Challenges and Opportunities

Indigenous and local communities often face significant challenges in gaining formal recognition of their land and resource rights. Disputes over land use, legal barriers, and the lack of official recognition can limit their ability to manage and protect their environments effectively. These challenges are compounded by insufficient access to financial resources and essential infrastructure, which can restrict the ability of these communities to implement and sustain environmental initiatives. Without adequate funding, technology, and support, it becomes difficult for Indigenous and local communities to address environmental challenges, especially as they are often disproportionately affected by climate change and environmental degradation (Hall, 1999; Eddy, 2010; Whittington, 2023).

On the other hand, there are opportunities to empower these communities through capacity-building efforts and the provision of resources. By investing in training, technical support, and funding, Indigenous and local communities can be better equipped to lead environmental management initiatives and implement sustainable solutions. Furthermore, integrating traditional knowledge systems with modern scientific approaches presents a significant opportunity for more comprehensive and effective environmental strategies. Collaboration between Indigenous knowledge holders and scientists can create innovative solutions that not only respect cultural heritage but also enhance conservation efforts, resulting in better outcomes for biodiversity and ecosystem protection.

Indigenous communities and local actors are vital to environmentalism transformation, bringing traditional knowledge, cultural connections, and community-based approaches to sustainability. Their involvement enriches environmental management practices, ensures inclusive and relevant policies, and contributes to effective conservation and resource management. By addressing challenges and leveraging opportunities, Indigenous and local stakeholders can drive significant progress in environmental protection and sustainability, fostering a more equitable and effective approach to global environmental challenges.

4.3. Intermediaries and Multipliers in Environmental Innovation

Intermediaries and multipliers play critical roles in environmental innovations' dissemination, scaling, and effectiveness. These entities bridge gaps between various stakeholders, facilitate knowledge transfer, and amplify the impact of sustainable solutions.

4.3.1. The Role of Intermediaries

Intermediaries act as facilitators between innovators and end-users, helping to transfer knowledge, technology, and practices. They can include a range of organizations and entities, such as research institutions, consultants, non-governmental organizations (NGOs), and industry associations (Qu, 2014; Pena, 2017).

Universities and research institutions play a pivotal role in knowledge transfer and dissemination by conducting studies, developing new technologies, and translating research into practical applications. These institutions serve as a bridge between scientific discovery and real-world implementation, sharing their findings with businesses, policymakers, and the general public. Environmental consultancy firms also contribute to knowledge dissemination by providing specialized expertise and guidance on sustainable practices, technologies, and regulatory compliance. They help businesses and governments effectively apply innovations, ensuring environmental regulations are met while promoting sustainability (Janssen et al., 2014; De Silva et al., 2022; Hermann et al., 2016).

Intermediaries also engage in capacity building and training efforts to equip stakeholders with the skills and knowledge needed to manage environmental innovations. They offer training programs, such as workshops, seminars, and certification courses, aimed at empowering individuals and organizations to implement sustainable solutions. Additionally, intermediaries provide technical assistance to help communities and businesses adopt new technologies and practices. This support often includes hands-on help, troubleshooting, and ongoing guidance to ensure successful implementation.

Networking and collaboration are further enhanced by intermediaries, who facilitate partnerships among innovators, businesses, governments, and other stakeholders. They organize conferences, forums, and networking events to foster connections and encourage the exchange of ideas. Intermediaries often create or engage in multi-stakeholder platforms that collaboratively bring together diverse groups to address environmental challenges. These platforms promote dialogue, share best practices, and develop joint solutions, encouraging a collective approach to sustainability.

4.3.2. The Role of Multipliers

Multipliers amplify the impact of environmental innovations by extending their reach and encouraging widespread adoption. They can include influential organizations, media, advocacy groups, and community leaders.

Media outlets and journalists play a major role in advocacy and promotion by raising awareness about environmental innovations and their benefits. Their reporting informs the public and policymakers, generating interest and support for new solutions. In parallel, environmental advocacy groups actively campaign to promote these innovations and influence policy changes. They mobilize public support, pressure governments and businesses, and highlight successful innovations to create a more favorable environment for sustainable practices (Thompson et al., 1990; Cardenete and Sancho, 2012; Zuhdi, 2014).

Once an environmental innovation succeeds in a pilot project, scaling and replication become critical. Multipliers can assist in expanding the reach of these innovations by showcasing their benefits and facilitating their adoption in other regions or sectors. This process often involves adapting the innovation to fit different contexts, ensuring it is relevant and effective in various settings. Funding and investment play a crucial role in this scaling process, as investors and funding organizations provide the necessary financial support to bridge the gap between initial development and large-scale implementation, thereby ensuring the sustainability of successful innovations.

Community engagement and mobilization are essential for the successful adoption of environmental innovations. Local leaders and organizations can champion these innovations within their communities, facilitating their integration into local practices. By tailoring solutions to meet the community's specific needs, they help build local support. Additionally, grassroots movements and activists play a significant role in driving the adoption of environmental innovations. These groups create a groundswell of support that encourages broader implementation of sustainable solutions by raising awareness, organizing campaigns, and mobilizing community action.

4.3.3. Challenges and Opportunities

Effective coordination and communication represent significant challenges for intermediaries and multipliers, particularly when navigating diverse stakeholders and complex environmental issues. Ensuring clear communication and alignment of goals is essential to foster collaboration and facilitate progress. Additionally, these entities often face resource constraints, including limitations in funding, capacity, and other essential resources. Such constraints can hinder their ability to effectively support and amplify environmental innovations (Van Oss and Padovani, 2003; Jansen and Vellema, 2004).

Despite these challenges, there are numerous opportunities for improvement. Strengthening collaborations among intermediaries, multipliers, and other stakeholders can enhance the dissemination and scaling of environmental innovations. By building robust partnerships and networks, these organizations can increase their impact and drive meaningful change. Furthermore, advancements in digital technology—such as social media, online platforms, and data analytics—offer new avenues for intermediaries and multipliers. These tools enable them to reach wider audiences, facilitate knowledge sharing, and effectively track progress in their initiatives, ultimately enhancing their capacity to address environmental challenges.

Intermediaries and multipliers are essential for advancing environmental innovation by bridging gaps between innovators and end-users, facilitating knowledge transfer, and amplifying the impact of sustainable solutions. Their roles in knowledge dissemination, capacity building, advocacy, and scaling are crucial for driving widespread adoption and achieving environmental goals. By addressing challenges and leveraging opportunities, these stakeholders can significantly contribute to the success and effectiveness of environmental innovations.

Chapter 5

Pathways for Environmental Transformation

Environmental transformation signifies a fundamental shift in societal attitudes and practices regarding managing natural resources, the approach to environmental challenges, and the pursuit of sustainable futures. This transformation is necessary as societies grapple with escalating environmental crises such as climate change, biodiversity loss, and pollution, all of which demand urgent and effective responses (Nesmith et al., 2021; Abbas et al., 2020; Yang et al., 2021). A multifaceted approach is essential to achieving this transformation. One pathway involves transitioning to renewable energy sources, which can significantly reduce greenhouse gas emissions while promoting energy security and resilience. Investments in solar, wind, and other sustainable energy technologies address climate change and stimulate job creation and economic growth.

Another critical pathway is adopting circular economy principles, prioritizing reducing waste and reusing resources. This approach challenges the traditional linear economy model, advocating for sustainable production and consumption patterns that minimize environmental impact. Additionally, local and indigenous knowledge can play a pivotal role in sustainable land management practices, fostering community resilience and enhancing biodiversity conservation.

Addressing environmental issues requires integrating social equity into environmental policies. Ensuring marginalized communities have a voice in decision-making can lead to more inclusive and effective solutions, as seen in grassroots movements advocating for environmental justice.

Environmental transformation requires collaboration among governments, businesses, and civil society to develop innovative solutions that are socially, economically, and ecologically sustainable. By exploring and implementing diverse pathways, societies can create a more resilient future that harmonizes human activity with the planet's health.

5.1. Pathways for Environmental Transformation

5.1.1. Policy and Regulatory Pathways

Strengthening environmental regulations is crucial for fostering sustainable practices and protecting natural resources. Developing and enforcing stricter legislation can significantly reduce pollution and ensure businesses adhere to sustainable practices. Key regulatory areas include emissions controls, waste management guidelines, and resource extraction standards. Additionally, implementing economic instruments—such as carbon pricing, renewable energy subsidies, and penalties for environmental violations—can provide the necessary incentives for businesses and individuals to adopt more sustainable practices (Liberti et al., 2016; Fealy et al., 2009; Bollyky et al., 2010).

An integrated policy approach is essential for addressing environmental sustainability comprehensively. This involves cross-sectoral integration, where environmental considerations are woven into policies across various sectors, including agriculture, energy, and transportation. Policymakers can create more effective solutions by ensuring that sustainability is a central focus in all aspects of development. Moreover, holistic planning is necessary to tackle multiple interrelated issues such as climate change, biodiversity loss, and resource depletion. Developing comprehensive environmental plans that address these challenges can promote coordinated actions and foster a more sustainable future.

5.1.2. Technological and Innovation Pathways

One of the most effective pathways to achieving environmental transformation is the advancement of green technologies. Promoting the development and adoption of renewable energy sources such as solar, wind, and hydropower can significantly reduce reliance on fossil fuels, lowering greenhouse gas emissions and combating climate change. By investing in these sustainable energy technologies, societies can transition toward cleaner energy systems that contribute to environmental health and foster economic growth and energy independence (Robinson et al., 2013; Hall et al., 2014; Linares et al., 2019).

In addition to renewable energy, enhancing energy efficiency is crucial in minimizing energy consumption and reducing overall environmental impact.

This can be achieved by investing in energy-efficient technologies and practices, including improved building designs, efficient appliances, and optimized industrial processes. Such innovations reduce the ecological footprint of various sectors and lead to cost savings for consumers and businesses alike.

Supporting research and development (RandD) is another vital component of environmental innovation. By focusing on sustainable materials, such as biodegradable plastics, eco-friendly construction materials, and advanced recycling technologies, RandD can drive significant advancements in product design and utilization. These innovations can help minimize waste and promote a circular economy.

Developing technologies that enhance climate resilience is essential for communities facing the impacts of climate change. Innovations such as early warning systems, drought-resistant crops, and effective flood management systems can empower communities to adapt to shifting environmental conditions. Through targeted research and development efforts, societies can build resilience and sustainability into their fabric, ensuring a more secure future for all.

5.1.3. Economic and Financial Pathways

Sustainable finance plays a crucial role in promoting environmental sustainability. We can direct capital towards sustainable initiatives by encouraging investment in green projects and technologies through green bonds, impact investing, and ESG criteria. Promoting corporate sustainability practices, including responsible supply chain management and environmental reporting, can also drive businesses to adopt more sustainable operations (Bardoscia et al., 2017; Xu et al., 2021; König et al., 2015).

The circular economy is another key component of a sustainable future. By adopting circular economic principles, such as reducing waste, recycling materials, and designing for durability, we can minimize resource consumption and environmental impact. Implementing practices that extend the life of products, such as repair and refurbishment services, can further reduce the need for new resources and decrease waste.

5.1.4. Social and Cultural Pathways

Integrating environmental education into school curricula and public awareness campaigns is essential for fostering a culture of sustainability. By equipping individuals with knowledge about their environmental impact, we can cultivate a generation that values ecological stewardship. Educational initiatives enhance awareness and empower individuals to make informed choices that contribute to environmental conservation. Additionally, engaging communities in environmental initiatives and decision-making processes ensures that local knowledge and perspectives are incorporated into solutions. This collaborative approach enhances the relevance and effectiveness of environmental strategies, creating a sense of ownership and responsibility among community members.

Promoting cultural shifts toward sustainable lifestyles is crucial for driving collective action for environmental change. Encouraging practices such as reducing consumption, supporting local products, and adopting eco-friendly habits can significantly positively impact the environment. By fostering a culture that values sustainability, we can inspire individuals and communities to make choices that benefit both people and the planet. Furthermore, recognizing and incorporating Indigenous knowledge and practices in environmental management can enhance sustainability efforts (Greenfield et al., 2003; Van der Brugge and Roosjen, 2015; de Royston et al., 2020). Indigenous communities possess valuable insights and traditional ecological knowledge that have allowed them to live in harmony with nature for generations. By respecting and integrating these practices, we honor cultural heritage and strengthen our collective efforts to address contemporary environmental challenges.

5.1.5. Institutional and Governance Pathways

Strengthening institutions and fostering collaboration are essential for effective global environmental governance. Building the capacity of environmental institutions and agencies to implement and enforce environmental policies effectively can improve governance and regulatory outcomes. Ensuring transparency and accountability in environmental governance, including public access to information and participation in decision-making, can enhance trust and effectiveness (Leach et al., 2007; Lassa, 2019; Lindahl et al., 2017).

Fostering partnerships between governments, businesses, NGOs, and communities can facilitate collaborative approaches to environmental challenges and leverage diverse expertise and resources. Engaging in international agreements and cooperation, such as the Paris Agreement and the Convention on Biological Diversity, can address transboundary environmental issues and promote global environmental goals.

5.1.6. Health and Well-Being Pathways

Conducting health impact assessments for environmental policies and projects is essential for ensuring that public health considerations are integrated into decision-making processes. By evaluating how proposed policies may affect community health, stakeholders can identify potential risks and benefits, allowing for more informed and responsible choices. This proactive approach helps prioritize health alongside environmental sustainability, fostering a comprehensive strategy that benefits both the environment and public health (Dilts et al., 2012; Ryff et al., 2002; Karunamun et al., 2021).

Promoting Healthy Environments Creating and maintaining healthy environments is vital for supporting overall well-being and quality of life. Access to clean air, safe drinking water, and green spaces play a crucial role in promoting physical and mental health. By prioritizing the establishment of healthy environments, policymakers can contribute to the prevention of illness, enhance community resilience, and improve the overall quality of life for residents.

5.1.6.1. Addressing Environmental Justice

Equity and Inclusion Ensuring that environmental policies and initiatives address the needs of marginalized and vulnerable communities is fundamental to promoting social equity and environmental justice. By recognizing the disproportionate impacts that environmental degradation can have on these populations, policymakers can develop targeted strategies that provide support and resources to those most affected. This commitment to equity not only fosters justice but also strengthens community resilience.

Participatory Approaches Involving affected communities in environmental decision-making and planning processes is crucial for addressing disparities and ensuring that solutions are fair and inclusive. Participatory approaches allow community members to share their experiences, knowledge, and concerns, leading to more relevant and effective policy outcomes. By actively engaging those impacted by environmental issues, decision-makers can develop solutions that reflect the diverse needs of all stakeholders, thereby fostering a more just and equitable approach to environmental management.

Pathways for environmental transformation encompass a wide range of strategies and approaches, including policy and regulatory reforms, technological advancements, economic incentives, social and cultural shifts, institutional strengthening, and health and well-being considerations. By exploring and implementing these pathways, societies can drive meaningful and sustained progress towards environmental sustainability and resilience, addressing the planet's complex and interconnected challenges.

5.2. Sustainable Development Goals and Environmentalism

The Sustainable Development Goals (SDGs), adopted by the United Nations in 2015, represent a global framework for addressing a wide range of critical issues, including environmental sustainability. The SDGs offer a comprehensive approach to achieving sustainable development by integrating economic, social, and environmental dimensions (Davey, 2009; Adams, 2005; Wapner, 2003).

5.2.1. Overview of the SDGs

The SDGs consist of 17 goals with 169 targets, designed to address global challenges and promote prosperity while protecting the planet. Several of these goals are directly related to environmentalism, focusing on issues such as climate change, biodiversity, clean water, and sustainable consumption (Roy et al., 2023).

5.2.2. Key Environmental SDGs

The Sustainable Development Goals (SDGs) are a set of 17 interconnected global goals aimed at promoting sustainable development and addressing global challenges. Several of these goals directly address environmental sustainability (Bernstein, 2001).

- *Goal 6:* Clean Water and Sanitation: Sustainable water management is crucial for maintaining ecosystems, supporting biodiversity, and ensuring access to clean water for all communities. Addressing water pollution and ensuring the efficient use of water resources are key aspects of environmental conservation.
- *Goal 7:* Affordable and Clean Energy: Transitioning to renewable energy sources reduces greenhouse gas emissions and mitigates climate change. Promoting energy efficiency also helps to lower environmental impacts and resource consumption.
- *Goal 12*: Responsible Consumption and Production: Responsible consumption and production are essential for reducing environmental degradation, conserving resources, and minimizing waste. This goal supports the transition to a circular economy and encourages sustainable practices across various sectors.
- *Goal 13*: Climate Action: Addressing climate change is critical for environmental sustainability. Mitigating greenhouse gas emissions, enhancing climate resilience, and adapting to climate impacts are central to protecting ecosystems and human communities.
- *Goal 14*: Life Below Water: Healthy marine environments are vital for biodiversity, climate regulation, and the livelihoods of coastal communities. This goal emphasizes the need for conservation efforts to protect marine life and ecosystems.
- *Goal 15:* Life on Land: Land ecosystems support a wide range of species and provide essential services such as carbon sequestration, water filtration, and soil fertility. This goal highlights the importance of protecting and restoring terrestrial environments.

These SDGs, along with other related goals, provide a comprehensive framework for addressing environmental challenges and promoting sustainable development.

5.2.3. The Integration of SDGs and Environmentalism

The interconnectedness of the Sustainable Development Goals (SDGs) creates opportunities for synergies and co-benefits. For instance, improving access to clean water (Goal 6) can enhance public health (Goal 3) by reducing waterborne diseases and contribute to sustainable agriculture (Goal 2) by ensuring adequate water supply for crops.

Addressing environmental challenges through the SDGs can lead to multiple benefits beyond the specific goals. For example, reducing poverty (Goal 1) can be achieved through sustainable development practices that create jobs and improve livelihoods. Promoting gender equality (Goal 5) can be facilitated by ensuring access to education and healthcare, which are often linked to sustainable development goals (Nilsson and Persson, 2017; Schleicher et al., 2018; Elder and Olsen, 2019).

5.2.3.1. Monitoring and Reporting

Effective monitoring and reporting are crucial for tracking progress towards the SDGs and ensuring accountability. Robust data and indicators are necessary to assess the effectiveness of policies and interventions. This includes measuring environmental indicators such as air quality, water pollution, and biodiversity loss.

Governments, organizations, and stakeholders must report on their progress towards achieving the SDGs. This involves publishing national reports, participating in global reviews, and sharing best practices and lessons learned. Transparent reporting helps to identify areas where progress is being made and where additional efforts are needed.

5.2.3.2. Stakeholder Engagement

Achieving the SDGs requires a collaborative approach that involves diverse stakeholders. Governments, businesses, civil society organizations, and communities must work together to develop and implement sustainable solutions. Engaging diverse stakeholders ensures that solutions are inclusive, equitable, and effective.

While the SDGs provide a global framework, local actions are essential for implementation. Communities and local organizations play a crucial role in driving sustainable practices and addressing specific environmental challenges. By involving local stakeholders, it is possible to tailor solutions to local needs and circumstances.

5.2.4. Challenges and Opportunities

Adequate funding and resources are critical for implementing and achieving the SDGs, especially in developing countries and vulnerable regions.

Financial constraints can hinder progress and limit the effectiveness of environmental initiatives. Investing in sustainable development requires significant financial commitments from governments, businesses, and international organizations.

Complex Interactions: The interconnected nature of the SDGs means that addressing environmental issues requires balancing multiple objectives and managing trade-offs. For example, promoting economic growth may sometimes conflict with environmental protection. This complexity can present challenges in policy and implementation.

5.2.4.1. Opportunities

Innovation and Technology: The SDGs provide an opportunity to drive innovation and technological advancements in renewable energy, waste management, and water conservation. Leveraging technology can enhance environmental sustainability and accelerate progress towards achieving the goals. For example, advancements in renewable energy technologies can help reduce greenhouse gas emissions and promote clean energy access.

Global Partnerships: The SDGs foster international cooperation and partnerships, enabling countries and organizations to share knowledge, resources, and best practices. Collaborative efforts can enhance the effectiveness and impact of environmental initiatives. By working together, countries can address global challenges such as climate change, biodiversity loss, and pollution more effectively.

The Sustainable Development Goals (SDGs) offer a comprehensive framework for addressing environmental challenges and promoting sustainability. By integrating environmental considerations into a broader development agenda, the SDGs provide pathways for achieving long-term environmental and socio-economic outcomes (Yen and Dung, 2023; Galanis, 2023; Morão and Visser, 2022). Addressing the key environmental SDGs, leveraging synergies, and engaging diverse stakeholders are crucial for advancing environmental transformation and ensuring a sustainable future for all.

5.3. Green Economy and Circular Economy Models

The Green Economy and Circular Economy models are two interrelated approaches aimed at achieving sustainable development by addressing environmental, economic, and social challenges. They offer frameworks for transforming economic activities and systems to align with ecological sustainability and resource efficiency (Melece, 2016; D'amato and Korhonen, 2021; Loiseau et al., 2016).

5.3.1. Green Economy

A Green Economy is one that prioritizes improving human well-being, social equity, and environmental sustainability. It aims to create an economy that is resource-efficient, low-carbon, and inclusive. Key principles of a Green Economy include sustainable development, reduced carbon emissions, resource efficiency, and social inclusion. By integrating environmental considerations into economic policies and practices, we can foster long-term sustainability.

A Green Economy emphasizes the adoption of low-carbon technologies to reduce greenhouse gas emissions. This includes promoting renewable energy sources, energy-efficient technologies, and cleaner production processes. Additionally, resource efficiency is crucial in a Green Economy. By promoting the efficient use of natural resources, we can minimize waste and environmental impact. Sustainable industries are also essential, encouraging businesses to adopt eco-friendly manufacturing processes, green supply chains, and sustainable agriculture (Georgeson et al., 2017; Victor and Jackson, 2012; Caprotti and Bailey, 2014).

A Green Economy offers numerous benefits. It reduces environmental degradation by promoting practices that limit pollution and conserve natural resources. It also supports economic growth by creating green jobs, fostering innovation, and opening new markets for green technologies and services. Furthermore, a Green Economy aims to address social inequalities by ensuring that the benefits of economic growth are distributed fairly and that marginalized communities are included in the transition.

However, transitioning to a Green Economy can present challenges. The transition may involve significant upfront costs for businesses and governments, which can be a barrier to implementation. Strong policy frameworks and regulatory measures are also necessary to support green initiatives and ensure compliance.

5.3.2. Circular Economy

A Circular Economy is an economic model that prioritizes minimizing waste, maximizing resource use, and enabling the continual reuse and recycling of materials. This contrasts with the traditional linear economy model of "take, make, dispose." Key principles of a Circular Economy include designing for longevity, promoting the reuse and recycling of materials, and creating closed-loop systems where waste from one process becomes input for another (Corvellec et al., 2022; Bhandari, 2023; Korhonen et al., 2018).

Product design plays a crucial role in a Circular Economy. Designing products for durability, repairability, and recyclability extends their lifecycle and reduces waste. Implementing systems for recovering valuable materials from used products, such as recycling and upcycling, is another key component. Additionally, adopting business models that support the Circular Economy, such as product-as-a-service and take-back schemes, can further promote sustainability.

The Circular Economy offers numerous benefits. It minimizes waste generation and environmental pollution by promoting the reuse and recycling of materials. It also enhances the efficiency of resource use by keeping materials in circulation and reducing the need for virgin resources. Furthermore, the Circular Economy creates new economic opportunities through innovative business models, waste management technologies, and recycling industries.

However, transitioning to a Circular Economy requires a systemic shift in how products are designed, manufactured, and consumed. This can be complex and resource intensive. Changing consumer behavior and attitudes towards product ownership and disposal can also be challenging. Despite these challenges, the Circular Economy offers a promising approach to sustainable development and a more resource-efficient future.

5.3.3. The Intersection of Green and Circular Economies

The Green and Circular Economies are complementary models that both aim to promote sustainability and reduce environmental impact. The Green Economy focuses on creating a low-carbon and resource-efficient economy, while the Circular Economy seeks to close material loops and reduce waste (Hoffman et al., 2022; Rask, 2022; Kumar et al., 2023). By integrating these two approaches, we can enhance the effectiveness of sustainability efforts.

For example, incorporating circular economy principles into green building design can enhance resource efficiency and reduce environmental impact. Using recycled materials and designing buildings for disassembly can minimize waste and promote the reuse of resources. Similarly, designing products that are both energy-efficient (green economy) and easily recyclable (circular economy) can lead to more sustainable consumption patterns.

The integration of Green and Circular Economies offers numerous benefits, including:

- *Enhanced sustainability:* Combining the principles of both models can lead to more comprehensive and effective environmental protection.
- *Increased resource efficiency*: By integrating circular economy principles into green technologies and industries, we can optimize resource use and reduce waste.
- *Reduced environmental impact*: The combination of low-carbon practices and circular economy principles can significantly reduce greenhouse gas emissions and environmental pollution.
- *Economic opportunities*: The integration of Green and Circular Economies can create new business opportunities and drive economic growth.
- *Social benefits*: By promoting sustainability and reducing environmental impact, the integration of Green and Circular Economies can contribute to improved public health, social equity, and overall well-being.

5.3.4. Implementation Strategies

5.3.4.1. Policies and Regulations

Supporting frameworks are essential for driving the adoption of green and circular economy practices. Developing and implementing policies that incentivize green technologies and regulate waste management and recycling can create a favorable environment for sustainable businesses and practices. Public-private partnerships can also play a crucial role in driving innovation and implementing sustainable solutions. By fostering collaboration between governments, businesses, and other stakeholders, we can leverage diverse expertise and resources to accelerate progress.

5.3.4.2. Innovation and Technology

Investing in research and development is crucial for advancing green technologies and circular economy solutions. By supporting innovation in areas such as new recycling methods and sustainable materials, we can develop more effective and efficient practices. Leveraging technology can also enhance resource efficiency. Advanced recycling technologies and smart resource management systems can help optimize resource use and reduce waste.

5.3.4.3. Education and Awareness

Providing training and resources to businesses and communities is essential for supporting the transition to green and circular economy practices. Capacity-building initiatives can help businesses adopt sustainable practices and improve their environmental performance. Raising public awareness about the benefits of green and circular economies is also crucial. By encouraging sustainable behaviors among consumers, we can create a market demand for green and circular products and services.

The Green Economy and Circular Economy models offer complementary approaches to achieving sustainability and addressing environmental challenges. By integrating these models, societies can foster more resilient and resource-efficient systems that support long-term environmental, economic, and social well-being (Khan et al., 2022; Barna et al., 2023; Gopalakrishnan et al., 2021). Implementing these approaches requires coordinated efforts across policy, technology, business practices, and consumer behavior to drive meaningful and sustained progress.

5.4. Renewable Energy Transitions

Renewable energy transitions involve shifting from fossil fuels and non-renewable energy sources to renewable sources like solar, wind, hydro, geothermal, and biomass. This transition is crucial for addressing climate change, improving energy security, and promoting sustainable development (York and Bell, 2019; Li et al., 2020; Hoicka et al., 2021).

5.4.1. Types of Renewable Energy

Renewable energy encompasses various sources that naturally replenish over time. Key types include solar energy, harnessed through photovoltaic systems and solar thermal technologies to generate electricity and heat. Wind energy is captured via onshore and offshore turbines, producing power from wind currents. Hydropower utilizes flowing water, primarily from rivers and dams, for electricity generation, while smaller systems like micro-hydropower minimize environmental impact. Geothermal energy taps into the Earth's internal heat for power and heating. Finally, biomass energy involves burning organic materials or converting them into biofuels, offering sustainable alternatives to fossil fuels for heat and transportation (Best, 2017; Sutherland et al., 2015).

5.4.1.1. Solar Energy

Solar energy harnesses sunlight through two primary systems: photovoltaic (PV) systems and solar thermal systems. Photovoltaic systems convert sunlight directly into electricity using solar cells. These systems can be installed on rooftops, in dedicated solar farms, or integrated into building materials for a seamless energy solution. In contrast, solar thermal systems use sunlight to heat fluids, producing steam that drives turbines for electricity generation or providing direct heating for residential and industrial applications.

5.4.1.2. Wind Energy

Wind energy is captured through onshore and offshore wind systems. Onshore wind involves the installation of wind turbines on land to generate electricity, typically in areas with high wind speeds. Onshore wind farms take advantage of these favorable conditions to produce renewable energy efficiently. Offshore wind energy, on the other hand, involves turbines installed in bodies of water, such as oceans or large lakes. These offshore wind farms can capture higher and more consistent wind speeds compared to their onshore counterparts, leading to increased energy production.

5.4.1.3. Hydropower

Hydropower utilizes the energy of flowing water to generate electricity, primarily through conventional hydropower systems. This method often involves dams or river systems and is one of the oldest and most established forms of renewable energy. Additionally, smaller-scale options like run-of-river and micro-hydropower systems do not require large dams and tend to have a lower environmental impact, making them suitable for various applications.

5.4.1.4. Geothermal Energy

Geothermal energy taps into the heat from the Earth's interior to generate electricity through geothermal power plants. This can be achieved using steam from hot water reservoirs or by injecting fluids into geothermal reservoirs. Another aspect of geothermal energy is the use of geothermal heat pumps, which leverage the stable temperatures of the Earth's surface for efficient heating and cooling of buildings.

5.4.1.5. Biomass Energy

Biomass energy involves the combustion of organic materials, such as wood, agricultural residues, or dedicated energy crops, to produce heat or electricity. In addition to direct combustion, biomass can also be converted into biofuels—liquid fuels like ethanol or biodiesel derived from organic materials. These biofuels serve as alternatives to fossil fuels and can be used in transportation, contributing to a more sustainable energy landscape.

5.4.2. Benefits of Renewable Energy Transitions

5.4.2.1. Environmental Benefits

Renewable energy sources offer significant environmental benefits, primarily by reducing greenhouse gas emissions. Unlike fossil fuels, which release large quantities of greenhouse gases into the atmosphere, renewable energy systems produce little to no emissions. This reduction plays a crucial role in mitigating climate change and its associated impacts. Additionally, renewable energy systems contribute to lower pollution levels, resulting in cleaner air and water compared to conventional energy sources (Bhattarai et al., 2022; Davidson, 2019). This shift not only supports ecological health but also helps preserve natural resources for future generations.

5.4.2.2. Economic Benefits

The economic advantages of renewable energy are substantial. The sector is a significant source of job creation, generating employment opportunities in various areas, including manufacturing, installation, maintenance, and research and development. These jobs not only provide livelihoods but also stimulate local economies. Furthermore, renewable energy contributes to energy independence by reducing reliance on imported fossil fuels. This enhanced national energy security fosters a more stable economic environment and promotes sustainable growth.

5.4.2.3. Social Benefits

The social benefits of renewable energy are equally important. By decreasing air pollution from fossil fuel combustion, renewable energy can improve public health outcomes, including fewer respiratory and cardiovascular diseases. Additionally, renewable energy technologies, particularly off-grid solutions like solar home systems, provide energy access to remote and underserved communities. This access enhances quality of life and promotes equity, ensuring more people benefit from reliable and sustainable energy sources.

5.4.3. Challenges and Barriers

Renewable energy transitions face several challenges and barriers. Technical and operational issues include the intermittency of sources like solar and wind, which necessitate effective energy storage solutions and grid upgrades for reliable supply. Economic barriers stem from high upfront costs and market structures that often favor fossil fuels, hindering the competitiveness of renewables. Additionally, regulatory challenges, such as policy uncertainty and complex permitting processes, can deter investment and slow project development (Nwanekezie et al., 2022; Goddard and Farrelly, 2018). Addressing these obstacles is crucial for accelerating the adoption of renewable energy and achieving sustainable energy goals globally.

5.4.3.1. Technical and Operational Challenges

Technical and operational challenges present significant hurdles in the transition to renewable energy. One major issue is the intermittency of sources like solar and wind, which are not always available when demand is high. To address this, effective energy storage solutions are necessary to ensure a stable energy supply. Additionally, integrating renewable energy into existing energy grids can be complex. The variable nature of some renewable sources necessitates upgrades to grid infrastructure, which can be both challenging and costly.

5.4.3.2. Economic and Financial Barriers

Economic and financial barriers also impede the growth of renewable energy. The initial investment required for renewable energy systems can be substantial, which may deter potential adopters. However, it's important to note that costs have been decreasing over time and can often be offset by long-term savings in energy bills. Furthermore, market structures in certain regions continue to favor fossil fuels, as subsidies and incentives are still predominantly geared toward conventional energy sources. This imbalance can hinder the competitiveness of renewable energy, making it more difficult for these systems to gain traction in the market.

5.4.3.3. Regulatory and Policy Challenges

Regulatory and policy challenges further complicate the advancement of renewable energy. Policy uncertainty stemming from changes in government regulations can create an unpredictable environment for investors and developers in the sector. This uncertainty can lead to hesitancy in committing to long-term projects. Additionally, the process of obtaining permits and licenses for renewable energy projects can be both complex and time-consuming, especially for large-scale installations. This bureaucratic burden can slow down the deployment of renewable energy technologies, impacting overall progress toward sustainable energy goals.

5.4.4. Pathways to Accelerate Renewable Energy Transitions

Accelerating renewable energy transitions requires a multifaceted approach. First, implementing supportive policies such as feed-in tariffs and renewable portfolio standards can create a favorable investment environment. Second, fostering technological innovation in energy storage and smart grid systems enhances the reliability and integration of renewable sources. Third, establishing robust financial mechanisms, including green financing and public-private partnerships, can provide the necessary capital for project development. Finally, investing in capacity building through training programs and public awareness campaigns ensures a skilled workforce and broad community support. Together, these pathways can facilitate a swift and effective transition to sustainable energy systems (Kemfert et al., 2020; Tzankova, 2020; Ofélia de Queiroz et al., 2024).

5.4.4.1. Policies and Regulations

Supportive policies play a crucial role in fostering the growth of renewable energy. Implementing measures such as feed-in tariffs, power purchase agreements, and renewable portfolio standards can create a stable environment for investment and development. These policies ensure that renewable energy producers receive fair compensation for their electricity, making it more attractive to enter the market. Additionally, providing financial incentives and subsidies, such as tax credits and grants, can significantly reduce cost barriers for renewable energy projects. These financial support mechanisms enable developers to overcome initial investment hurdles, promoting broader adoption and accelerating the transition to sustainable energy sources.

5.4.4.2. Technological Innovation

Technological innovation is vital for addressing the challenges associated with renewable energy transitions. Significant advancements in energy storage technologies, such as batteries and pumped hydro storage, are essential for mitigating intermittency issues that arise from sources like solar and wind. By investing in research and development in these areas, we can improve the reliability of renewable energy systems. Furthermore, the development of smart grid technologies enhances the integration and management of renewable energy sources within existing energy grids. Smart grids enable real-time monitoring and optimization of energy distribution, facilitating a more efficient use of renewable resources and promoting grid resilience.

5.4.4.3. Financial Mechanisms

Robust financial mechanisms are essential for driving investment in renewable energy projects. Green financing options, such as green bonds, impact investing, and climate funds, can channel capital toward the development of sustainable energy initiatives. These financial instruments provide necessary funding for projects that might otherwise struggle to secure investment. Additionally, public-private partnerships play a critical role in facilitating the deployment and scaling of renewable energy technologies. By fostering collaboration between governments, businesses, and financial institutions, these partnerships can leverage resources, share risks, and enhance the overall effectiveness of renewable energy initiatives.

5.4.4.4. Capacity Building and Education

Capacity building and education are vital components of a successful renewable energy transition. Implementing training programs for the renewable energy workforce can ensure that skilled labor is available for the installation and maintenance of renewable technologies. This investment in human capital not only supports sector growth but also enhances job opportunities in emerging fields. Moreover, raising public awareness about the benefits and potential of renewable energy can build community support and drive adoption. By educating the public on the advantages of renewable energy, including its role in combating climate change and promoting energy independence, we can foster a more informed and engaged citizenry committed to sustainable practices.

5.4.5. Case Studies and Examples

5.4.5.1. Global Success Stories

Germany is a notable example of a successful renewable energy transition through its Energiewende (energy transition) policy. This comprehensive approach has led to substantial investments in renewable energy infrastructure, resulting in impressive growth in both solar and wind power capacities. Germany's commitment to sustainability has not only transformed its energy landscape but has also made it a model for other nations seeking to enhance their renewable energy initiatives.

Conversely, China has emerged as a global leader in renewable energy production and installation. The country has invested heavily in solar and wind power technologies, becoming the largest producer of renewable energy systems worldwide. China's aggressive expansion in this sector reflects its commitment to addressing climate change and reducing dependence on fossil fuels, positioning the nation as a pivotal player in the global renewable energy market.

5.4.5.2. Innovative Projects

Innovative projects also exemplify the advancements in renewable energy technology. The Tesla Powerwall is a prominent home battery storage system that allows homeowners to store solar energy for later use, particularly during periods when sunlight is unavailable. This technology not only enhances energy independence but also contributes to the overall efficiency of residential solar systems.

Another significant initiative is Desertec, a large-scale project designed to harness solar power from the deserts of North Africa. The aim is to generate renewable energy that can be exported to Europe, showcasing the potential of cross-border renewable energy solutions. This ambitious project highlights the capacity for innovation in renewable energy deployment, addressing both local energy needs and global sustainability goals.

Renewable energy transitions are essential for addressing climate change, enhancing energy security, and promoting sustainable development. While there are challenges to overcome, including technical, financial, and regulatory barriers, the benefits of renewable energy are substantial and far-reaching (Rosenbloom and Meadowcroft, 2022; Del Granado et al., 2020). By implementing supportive policies, advancing technologies, and fostering collaboration, societies can accelerate the transition to renewable energy and build a more sustainable and resilient energy future.

5.5. Biodiversity Conservation and Ecosystem Services

Biodiversity conservation and ecosystem services are critical to environmental sustainability and human well-being. Biodiversity encompasses the variety of life forms on Earth, including genetic, species, and ecosystem diversity. Ecosystem services refer to the benefits humans derive from ecosystems, which are essential for survival and quality of life (Polasky, 2006; Rands et al., 2010).

5.5.1. Biodiversity Conservation

5.5.1.1. Definition and Importance

Biodiversity conservation refers to the protection, preservation, and management of the variety of life on Earth. This effort aims to prevent the loss of species, habitats, and genetic diversity, ensuring that ecosystems remain robust and functional. The importance of biodiversity cannot be overstated; it is crucial for maintaining ecosystem health, resilience, and overall functioning. Biodiversity supports essential processes such as pollination, nutrient cycling, and climate regulation. Additionally, it provides vital resources for food, medicine, and materials that are fundamental to human survival and well-being (Reed et al., 2011; Paoli et al., 2010; Dawson et al., 2011).

5.5.1.2. Levels of Biodiversity

Biodiversity can be understood at three distinct levels. First, genetic diversity refers to the variation in genetic material within species, which is essential for adaptability and resilience in the face of environmental changes. Second, species diversity encompasses the variety of species within a specific ecosystem or region, contributing to ecosystem stability and functionality. Finally, ecosystem diversity highlights the range of different ecosystems present in a given area, including forests, wetlands, grasslands, and marine environments. Together, these levels of biodiversity form the foundation of healthy ecosystems.

5.5.1.3. Threats to Biodiversity

Despite its importance, biodiversity faces numerous threats. Habitat destruction caused by deforestation, urbanization, and land conversion leads to the loss and fragmentation of habitats, directly impacting species survival. Climate change further exacerbates these threats by altering habitats and ecosystems, causing shifts in species distributions, changes in breeding patterns, and increasing extinction risks. Additionally, overexploitation—through unsustainable harvesting of resources like fishing, hunting, and logging—can deplete species populations and disrupt ecosystems. Pollution from chemicals and plastics poses another significant threat, harming wildlife and reducing biodiversity. Furthermore, invasive species introduced to new environments can outcompete native species, alter habitats, and create ecological imbalances.

5.5.1.4. Conservation Strategies

To combat these threats, various conservation strategies are employed. Establishing protected areas, such as national parks, wildlife reserves, and marine protected areas, is vital for safeguarding critical habitats and species. Restoration projects aim to rehabilitate degraded ecosystems and restore native species, thereby improving habitat quality and connectivity. Sustainable management practices seek to balance human needs with environmental protection, encompassing sustainable forestry and fisheries management. Legal frameworks are essential for enforcing laws and regulations that prevent illegal wildlife trade, protect endangered species, and conserve critical habitats. Finally, community involvement is crucial for the success of conservation efforts; engaging local communities, incorporating traditional knowledge, and promoting stewardship of natural resources enhance the effectiveness of biodiversity conservation initiatives.

5.5.2. Ecosystem Services

Ecosystem services refer to the benefits that humans derive from ecosystems, which can be categorized into four main types: provisioning, regulating, supporting, and cultural services. Provisioning services encompass the tangible products obtained from ecosystems, such as food, fresh water, timber, and medicinal resources. Regulating services involve the benefits derived from the regulation of ecosystem processes, including climate regulation, flood control, water purification, and pollination. Supporting services are those that underpin the functioning of ecosystems, such as nutrient cycling, soil formation, and primary production. Lastly, cultural services provide non-material benefits obtained from ecosystems, offering recreational, aesthetic, spiritual, and educational values (Bennett et al., 2009; Carpenter et al., 2006; Wallace, 2007).

5.5.2.1. Examples of Ecosystem Services

Numerous examples illustrate the vital role of ecosystem services in supporting human well-being. Pollination is a crucial service, where insects and other animals pollinate crops and wild plants, essential for food production and overall ecosystem health. Water filtration is another significant service provided by wetlands and riparian zones, which filter pollutants from water, thus improving water quality and reducing the need for costly treatment. Additionally, forests and oceans contribute to climate regulation by sequestering carbon dioxide, which helps mitigate climate change and regulate global temperatures. Soil fertility is critical for agriculture, as healthy soils provide essential nutrients that enhance crop productivity. Moreover, natural landscapes and ecosystems offer cultural and recreational values, providing opportunities for recreation, tourism, and spiritual enrichment.

5.5.2.2. Challenges in Ecosystem Services

Despite their importance, ecosystem services face several challenges. Degradation and loss of ecosystems, such as deforestation and wetland loss, can significantly diminish their capacity to provide essential services. Furthermore, accurately valuing and accounting for ecosystem services can be challenging but is crucial for sustainable management. Incorporating these values into decision-making processes helps highlight their significance. Additionally, climate change impacts pose a threat, as alterations in climate can affect the delivery and reliability of ecosystem services, necessitating adaptive management strategies to address these challenges.

5.5.2.3. The Integration of Conservation and Ecosystem Services

Integrating conservation efforts with the management of ecosystem services is essential for ensuring sustainable outcomes. Ecosystem-based management offers a holistic approach that considers the interconnections between ecosystems and the services they provide. Natural capital accounting is another critical strategy, incorporating the value of ecosystem services into economic and policy decisions to emphasize their importance and support conservation efforts. Finally, the Sustainable Development Goals (SDGs) underscore the significance of conserving biodiversity and maintaining ecosystem services, with specific goals such as Goal 14 (Life Below Water) and Goal 15 (Life on Land) highlighting the need for integrated conservation strategy.

Biodiversity conservation and ecosystem services are deeply interconnected and essential for maintaining the health of our planet and human well-being. Effective conservation strategies that protect biodiversity also ensure the continued provision of ecosystem services that support life on Earth. Addressing threats to biodiversity and ecosystem services through integrated approaches, sustainable practices, and collaborative efforts is crucial for achieving long-term environmental sustainability and resilience.

5.6. Decarbonization and Carbon Neutrality Pathways

Decarbonization and carbon neutrality are critical goals in the fight against climate change, aiming to significantly reduce or eliminate greenhouse gas (GHG) emissions to mitigate global warming and its associated impacts (Perissi andJones, 2022; Zhang and Hanaoka, 2022; Lee et al., 2024).

5.6.1. Decarbonization

5.6.1.1. Definition and Objectives

Decarbonization is defined as the process of reducing carbon dioxide (CO_2) emissions and other greenhouse gases generated by human activities. This is primarily achieved through transitioning to cleaner energy sources and enhanced efficiency across various sectors. The overarching objectives of decarbonization are to limit global temperature rise, mitigate the impacts of climate change, and facilitate a transition toward a sustainable, low-carbon economy (Wang et al., 2023; Shao et al., 2022; Zhou et al., 2023).

5.6.1.2. Key Strategies

One of the key strategies for decarbonization is the transition to renewable energy. This involves increasing the share of energy derived from renewable sources such as solar, wind, hydro, geothermal, and biomass. Such a shift not only reduces reliance on fossil fuels but also significantly lowers carbon emissions. Additionally, modernizing energy grids is crucial to accommodate variable renewable energy sources and improve the efficiency of energy transmission and distribution.

Another important strategy is to enhance energy efficiency improvements across various sectors. In buildings—whether residential, commercial, or industrial—this can be achieved through better insulation, the use of energy-efficient appliances, and the implementation of smart energy management systems. In the transportation sector, promoting fuel-efficient vehicles, electric vehicles (EVs), and alternative fuels is essential. Additionally, improving public transportation systems and encouraging active modes of transportation, such as walking and cycling, can lead to significant emission reductions.

Industrial decarbonization is also a critical component of the overall strategy. This can be achieved by implementing cleaner technologies and practices in industrial processes, such as carbon capture and storage (CCS) and the use of low-carbon materials. Embracing circular economy principles can further reduce waste and enhance resource efficiency in industrial operations.

In the realms of agriculture and land use, promoting sustainable practices is vital. This includes encouraging methods such as conservation tillage, agroforestry, and improved livestock management to minimize emissions and enhance carbon sequestration in soils. Moreover, initiatives like reforestation and afforestation can significantly contribute to increasing forest cover and restoring degraded lands, thereby enhancing carbon sequestration.

Carbon pricing and market mechanisms are essential for incentivizing emission reductions. Implementing carbon taxes can internalize the cost of carbon emissions, motivating businesses to lower their emissions. Similarly, establishing cap-and-trade systems can create economic incentives for emissions reduction by capping overall emissions and allowing businesses to trade allowances.

Lastly, innovation and technology play a pivotal role in achieving decarbonization goals. Investing in research and development (RandD) for new technologies that support decarbonization—such as advanced energy storage and next-generation renewable technologies—is critical. Additionally, supporting the deployment and scaling of proven technologies can accelerate the transition to a low-carbon economy.

5.6.1.3. Challenges and Barriers

Despite the outlined strategies, decarbonization faces several challenges and barriers. One significant hurdle is the cost and investment associated with renewable energy infrastructure and energy efficiency upgrades, which can be particularly challenging for developing economies. Furthermore, existing infrastructure and technology limitations may impede the rapid deployment of new practices and technologies.

Inconsistent policy and regulation can also slow down progress, creating uncertainty for investors and businesses. A lack of cohesive regulatory support may hinder efforts to transition to a low-carbon economy, underscoring the need for stable and supportive policies to facilitate the decarbonization process.

5.6.2. Carbon Neutrality

Carbon neutrality is defined as the process of balancing the amount of carbon dioxide (CO_2) emitted with an equivalent amount of CO_2 removal or offsetting, ultimately resulting in a net-zero carbon footprint. Achieving carbon neutrality involves not only reducing emissions but also compensating for any remaining emissions through various offsetting measures. The primary objective is to attain a net-zero carbon footprint for organizations, countries, or individuals, thereby contributing to global climate goals and minimizing the overall impact on climate change (Williams et al., 2021; He et al., 2022).

5.6.2.1. Pathways to Achieving Carbon Neutrality

A crucial pathway to achieving carbon neutrality is emission reduction. This begins with quantifying emissions by measuring and monitoring emissions from all relevant sources, including energy consumption, transportation, and industrial processes. Setting clear and ambitious emission reduction targets aligned with global climate agreements and science-based benchmarks is vital. Subsequently, organizations can implement various strategies to reduce emissions, as previously outlined in the decarbonization section.

In addition to emission reductions, carbon offsetting plays an essential role. This involves investing in projects that actively remove or reduce greenhouse gases (GHGs) from the atmosphere, such as reforestation, afforestation, and methane capture from landfills. To ensure the credibility and effectiveness of these offsets, it is crucial to choose projects certified by recognized standards, such as the Verified Carbon Standard (VCS) or Gold Standard.

Another important strategy is the use of carbon capture and storage (CCS) technologies. These technologies capture CO_2 emissions from industrial processes and power plants before they are released into the atmosphere. Once captured, the CO_2 can be securely stored in geological formations or other safe storage methods to prevent it from re-entering the atmosphere.

Additionally, pursuing climate neutral certification can help organizations demonstrate their commitment to carbon neutrality. Obtaining certification from recognized programs or organizations verifies carbon neutrality and showcases efforts in emission reductions and offsets. Maintaining transparency in emissions reporting and offsetting activities is critical, as is regularly reviewing and updating strategies to ensure continued progress toward carbon neutrality.

5.6.2.2. Challenges and Barriers

Despite the clear pathways, several challenges and barriers must be addressed in the pursuit of carbon neutrality. One significant challenge is ensuring the verification and integrity of carbon offset projects and CCS solutions. This requires robust monitoring and verification processes to ensure accuracy and trustworthiness.

Another hurdle is the need for long-term commitment. Achieving and maintaining carbon neutrality requires ongoing dedication and the ability to adapt to changing technologies, regulations, and market conditions.

Equity and access to carbon offset opportunities and technologies must be considered, particularly for developing countries and marginalized communities. Ensuring equitable access is essential for fostering global fairness in the pursuit of carbon neutrality and addressing climate change effectively.

5.6.3. The Integration of Decarbonization and Carbon Neutrality

A holistic approach to tackling climate change involves integrating decarbonization strategies with carbon neutrality goals. This comprehensive framework encompasses various components, including emission reductions, carbon offsets, and technological innovations. By merging these elements, organizations and governments can create a unified strategy that effectively addresses the complexities of achieving net-zero emissions (Hao-Nan et al., 2022; Bauer et al., 2022; Monjardino et al., 2021).

Sectoral coordination is crucial for maximizing the impact of decarbonization and carbon neutrality initiatives. Efforts must be synchronized across multiple sectors—such as energy, transportation, industry, and agriculture—to achieve synergistic effects. By fostering collaboration and sharing best practices among these sectors, stakeholders can enhance overall outcomes and drive more significant progress toward shared climate objectives.

Additionally, policy alignment at both national and international levels is essential for supporting decarbonization and carbon neutrality goals. Creating policies that incentivize action and provide a supportive framework is necessary to facilitate the transition to a low-carbon economy. When governments align their regulations and initiatives with these objectives, they can better mobilize resources, attract investments, and encourage participation from all sectors of society. This coordinated approach is vital for ensuring the success of efforts aimed at combating climate change and achieving sustainable development.

Decarbonization and carbon neutrality are pivotal in addressing climate change and achieving long-term environmental sustainability. Societies can work towards a low-carbon future by implementing strategies to reduce emissions, investing in clean technologies, and supporting carbon offset projects. Overcoming challenges and barriers requires coordinated efforts, innovative solutions, and robust policies to ensure successful transitions and contribute to global climate goals.

Chapter 6

Environmental Knowledge Systems and Innovation

Environmental knowledge systems and innovation are crucial in addressing environmental sustainability's complex and evolving challenges. They encompass how knowledge about the environment is generated, shared, and utilized to drive solutions and advancements (Su et al., 2020; Fischer, 2001; Papaioannou et al., 2009).

6.1. Environmental Knowledge Systems

Environmental knowledge systems encompass the structured methodologies through which information and understanding about the environment are acquired, organized, and applied. These systems integrate various elements, including scientific research, traditional knowledge, policy frameworks, and public awareness initiatives. Together, they form a comprehensive approach to understanding and addressing environmental challenges (Stefik, 2014; Boiral, 2002; Gennari et al., 2003).

Scientific research serves as a critical component of environmental knowledge systems. It involves systematic investigation aimed at acquiring new knowledge or validating existing understanding related to environmental processes, impacts, and potential solutions. This empirical approach provides the foundation for informed decision-making and effective environmental management.

Traditional knowledge, on the other hand, represents the insights gained from indigenous and local knowledge systems. These perspectives offer valuable information about sustainable practices, ecological relationships, and resource management methods that have been developed over generations. By integrating traditional knowledge with scientific research, a more holistic understanding of environmental issues can be achieved.

Policy and governance frameworks play a vital role in shaping environmental knowledge systems. These frameworks and regulations guide environmental management and decision-making by leveraging scientific evidence and considering societal needs. Effective policies can enhance the application of environmental knowledge in practice, ensuring that decisions are grounded in sound understanding.

Education and public awareness initiatives are essential for fostering a broader understanding of environmental issues. By promoting sustainable behaviors among the public, these initiatives help to create an informed citizenry that can engage in environmental stewardship. Together, these components of environmental knowledge systems work synergistically to address the complexities of environmental challenges.

6.1.1. Research Institutions and Academia

Academic research is a cornerstone of environmental knowledge systems. Universities and research institutions conduct extensive studies on a wide array of environmental issues, including climate change, biodiversity conservation, and ecosystem management. The findings from these studies are often disseminated through peer-reviewed journals and scientific publications, which serve as key outputs that contribute to the collective understanding of environmental challenges and solutions (Stefik, 2014; Boiral, 2002; Gennari et al., 2003. This rigorous process ensures that the knowledge generated is credible and can be utilized by policymakers, practitioners, and the public.

Collaborative research further enhances the development of environmental knowledge by fostering partnerships among institutions, governments, and the private sector. These collaborations facilitate interdisciplinary research and innovation, allowing for diverse perspectives and expertise to be integrated into environmental studies. By combining resources and knowledge from various stakeholders, collaborative research can address complex environmental problems more effectively and develop innovative solutions that are informed by a broader understanding of the issues at hand. Such partnerships are essential for advancing research agendas that align with societal needs and environmental sustainability goals.

6.1.2. Indigenous and Local Knowledge

Indigenous and local knowledge plays a crucial role in understanding and managing ecosystems sustainably. Oral traditions, which encompass stories, practices, and knowledge passed down through generations, provide valuable insights into local ecosystems and sustainable management practices. These narratives encapsulate a deep understanding of the land, species, and ecological relationships that have been honed over centuries. Engaging indigenous and local communities through participatory approaches is essential for incorporating their knowledge into environmental management (Lam et al., 2020; Brondízio et al., 2021; Wheeler and Root-Bernstein, 2020). By actively involving these communities in research and decision-making processes, policymakers can ensure that local insights are respected and integrated into broader environmental strategies.

6.1.2.1. Policy and Legislation

Regulatory frameworks serve as the backbone of environmental management, guiding policies and laws that regulate environmental protection, resource management, and pollution control. These frameworks are informed by scientific evidence and stakeholder input, ensuring that diverse perspectives are considered in the decision-making process. Additionally, international agreements, such as the Paris Agreement and the Convention on Biological Diversity, provide global frameworks that guide collective action on pressing environmental issues. These agreements foster collaboration among nations and encourage shared responsibility in addressing global challenges.

6.1.2.2. Media and Communication

The media plays a vital role in disseminating knowledge about environmental issues. Environmental reporting through news media, documentaries, and online platforms raises awareness and educates the public about pressing challenges. Educational programs, including school curricula, public seminars, and workshops, are designed to inform individuals about environmental issues and potential solutions. By enhancing public understanding, these initiatives can inspire collective action and promote sustainable behaviors within communities.

6.1.2.3. Challenges in Environmental Knowledge Systems

Despite the wealth of knowledge available, several challenges persist within environmental knowledge systems. Integration and coordination of diverse sources of knowledge, such as scientific research and traditional knowledge, can be complex. The difficulty in merging these different approaches into cohesive and actionable strategies often hinders effective decision-making. Additionally, data gaps and uncertainties regarding environmental variables and processes pose significant barriers to understanding and addressing environmental challenges. Finally, ensuring equitable access to environmental knowledge and resources, particularly for marginalized communities and developing regions, remains a critical issue that must be addressed to promote fairness and inclusivity in environmental governance.

6.1.3. Innovation in Environmental Solutions

Innovation in environmental solutions encompasses the development and application of new technologies, practices, and approaches aimed at addressing environmental challenges and promoting sustainability. The significance of such innovations lies in their potential to lead to more effective and efficient methods for managing natural resources, reducing environmental impacts, and enhancing resilience to the effects of climate change (González-García et al., 2012; Biondi et al., 2002; Faivre et al., 2017). By fostering innovative practices, societies can transition towards sustainable futures while simultaneously addressing pressing ecological concerns.

6.1.3.1. Types of Environmental Innovation

Environmental innovation can be categorized into several key areas:

- *Technological Innovations*: These include advancements in clean energy technologies, such as solar panels, wind turbines, and energy storage systems. Additionally, pollution control technologies focus on improving air and water filtration, waste management, and emissions reduction.
- *Sustainable Practices*: Strategies like the circular economy promote the reuse, recycling, and reduction of waste materials, aiming to create a closed-loop system where products and materials are continuously cycled. Green building practices prioritize energy efficiency, sustainable materials, and overall environmental performance in construction.
- *Data and Monitoring Technologies*: Innovations such as remote sensing and Geographic Information Systems (GIS) provide real-time data and spatial analysis for monitoring environmental changes, including deforestation and climate impacts. Smart sensors and

Internet of Things (IoT) devices further enhance data collection and management capabilities for addressing environmental issues.

- *Policy and Governance Innovations*: Market-based instruments, including carbon pricing, green bonds, and environmental credits, incentivize sustainable practices and investments. Adaptive management approaches promote flexibility and iterative decision-making, allowing policies to be adjusted based on new information and changing conditions.

6.1.3.2. Promoting and Scaling Innovation

To effectively promote and scale environmental innovation, several strategies can be employed:

- *Research and Development*: Funding and investment are crucial for accelerating the development of new technologies and practices through grants, venture capital, and public funding. Establishing innovation hubs, including research centers, incubators, and accelerators, can provide essential resources and support for startups and researchers focused on environmental innovations.
- *Collaboration and Partnerships*: Public-private partnerships can facilitate collaboration between governments, businesses, and NGOs to develop and implement innovative solutions to environmental challenges. Additionally, international cooperation enables the sharing of knowledge, technologies, and best practices across countries and regions to tackle global environmental issues effectively.
- *Education and Capacity Building*: Training programs aimed at enhancing education and skills development for individuals and organizations involved in environmental management and innovation are vital. Awareness campaigns can also promote the benefits and importance of environmental innovation, increasing support and adoption among stakeholders and the public.

6.1.3.3. Challenges in Environmental Innovation

Despite the potential benefits, several challenges hinder environmental innovation. Implementation and adoption of new technologies and practices often face barriers, such as high costs, resistance to change, and regulatory hurdles. Additionally, ensuring that innovative solutions are scalable and have a meaningful impact on environmental outcomes is critical. Finally, addressing disparities in access to innovative solutions is essential for promoting equity and inclusion, ensuring that the benefits of innovation are distributed fairly across different communities and regions.

Environmental knowledge systems and innovation are essential for addressing environmental sustainability's complex and interconnected challenges. By integrating diverse sources of knowledge, supporting technological advancements, and fostering collaborative efforts, societies can develop effective solutions to protect natural resources, mitigate climate change, and promote resilience (González-García et al., 2012; Biondi et al., 2002; Faivre et al., 2017). Addressing challenges and leveraging opportunities in these areas will be crucial for achieving long-term environmental goals and ensuring a sustainable future for all.

6.2. Indigenous and Local Knowledge Systems

Indigenous and local knowledge systems represent a profound and diverse body of understanding about the environment, developed over centuries through lived experiences, cultural practices, and close interactions with nature. These systems provide valuable insights into sustainable resource management, ecosystem dynamics, and cultural heritage (Fernández-Llamazares et al., 2021; Smith et al., 2017; Lam et al., 2020).

6.2.1. Definition and Scope

Indigenous Knowledge refers to the traditional understanding held by Indigenous peoples and communities, often passed down through generations. This knowledge includes a deep connection to the environment, covering ecological relationships, resource management practices, and cultural values. Local Knowledge, on the other hand, pertains to the insights and practices of non-Indigenous local communities that have lived in a specific area for a long time. Like Indigenous Knowledge, this understanding is also rooted in long-term interactions with the environment and can overlap or complement Indigenous perspectives (Liporacci et al., 2015; Howes, 1979; Berkes et al., 2007).

The scope of Indigenous and Local Knowledge encompasses several key areas. Firstly, it includes ecological understanding, which involves knowledge of local ecosystems, species, seasonal changes, and environmental shifts. Secondly, resource management focuses on techniques and practices for sustainably harvesting and conserving natural resources. Additionally, cultural practices encompass traditions, rituals, and social norms related to the environment and its care. Finally, adaptation strategies refer to methods developed to cope with environmental changes, such as climate variability and natural disasters.

6.2.2. Characteristics of Indigenous and Local Knowledge

Indigenous and Local Knowledge adopts a holistic perspective, recognizing the interconnectedness of humans, nature, and spiritual beliefs. This means that environmental management is often intertwined with social and cultural practices. Additionally, systems thinking is integral to this knowledge, viewing the environment as a complex system where changes in one part can significantly affect the whole, encompassing ecological, social, and cultural dimensions (Klenk et al., 2017; Pawluk et al., 1992; Bang, 2016).

- *Place-Based Knowledge*: This type of knowledge is highly localized, specific to the geographic and ecological context of the community. It includes a detailed understanding of local flora, fauna, and landforms. Furthermore, it is experiential, rooted in direct experiences and observations over long periods. This long-term engagement leads to a deep understanding of local environmental dynamics.
- *Dynamic and Adaptive*: Indigenous and Local Knowledge is not static; it is evolutionary and evolves over time as communities adapt to environmental changes, new challenges, and external influences. This knowledge system is responsive and flexible, capable of incorporating new information and adjusting practices based on changing conditions.
- *Oral Tradition:* Storytelling is a vital method of transmitting knowledge within these communities. Oral traditions, including stories, songs, and rituals, encode important environmental information and cultural values. Elders and knowledge keepers play a crucial role in this process, preserving and passing on this wisdom to future generations.

6.2.3. Applications and Contributions

Sustainable resource management involves traditional practices that enhance environmental resilience and sustainability. Techniques such as rotational grazing, agroforestry, and community-managed fisheries are examples of methods that promote the sustainable use of resources. Additionally, conservation strategies, including sacred groves, taboo areas, and seasonal closures, help protect biodiversity and ecosystems, ensuring the long-term health of the environment (Kashif et al., 2023; Cairns, 2007; Brugnach et al., 2017).

6.2.3.1. Climate Adaptation and Resilience

Indigenous and local knowledge plays a vital role in climate adaptation and resilience. Communities possess an understanding of local climate patterns and employ adaptive strategies, such as adjusting planting times, constructing buildings that can withstand local weather conditions, and managing water resources effectively. Moreover, they implement risk mitigation practices to manage natural hazards, including flood management techniques and fire management strategies, helping to protect their communities from environmental challenges.

6.2.3.2. Cultural Heritage and Identity

Traditional ecological knowledge is deeply intertwined with cultural heritage and identity, providing a sense of belonging and continuity for communities. This knowledge not only informs sustainable practices but also imparts important values related to stewardship and respect for nature. Furthermore, it fosters community cohesion, reinforcing the connections between people and their environment.

6.2.3.3. Research and Policy Integration

Collaborative research efforts between indigenous and local communities and scientific researchers are essential for integrating traditional knowledge into environmental management. These partnerships facilitate the inclusion of indigenous perspectives in scientific research, enhancing the understanding of ecosystems. Additionally, advocating for the incorporation of indigenous and local knowledge into environmental policies, conservation strategies, and climate action plans is crucial for creating effective and inclusive approaches to environmental challenges.

6.2.4. Challenges and Opportunities

Indigenous and local knowledge faces several significant challenges. One major issue is ensuring that this knowledge is recognized and respected within broader environmental and scientific communities. Without proper acknowledgment, valuable insights may be overlooked or disregarded. Additionally, there is a pressing need to protect traditional knowledge from unauthorized use or appropriation. This includes ensuring fair and equitable benefit-sharing arrangements for the communities that hold this knowledge. Moreover, climate change poses substantial challenges by potentially disrupting traditional practices and necessitating adaptations within these knowledge systems to cope with changing environmental conditions (Ifejika Speranza et al., 2010; Johnson et al., 2016; Boillat and Berkes, 2013).

Despite these challenges, there are numerous opportunities to enhance the integration of indigenous and local knowledge into environmental management. One promising avenue is the development of enhanced collaboration between indigenous and local communities, scientists, and policymakers. By building partnerships, these groups can leverage their diverse knowledge systems for more effective environmental management. Furthermore, facilitating knowledge exchange programs can promote mutual learning and respect, allowing for the integration of traditional and scientific knowledge. Lastly, incorporating indigenous and local knowledge into policy frameworks and management plans can lead to more inclusive and effective solutions, ensuring that a wider range of perspectives and practices is considered in addressing environmental challenges.

6.2.5. Case Studies and Examples

6.2.5.1. Fire Management in Australia

In Australia, traditional burning practices utilized by Indigenous peoples are becoming increasingly recognized for their effectiveness in fire management. Controlled burns, which are conducted to reduce fuel loads, play a crucial role in preventing large-scale wildfires. By integrating these traditional practices with modern fire management strategies, Australia aims to enhance its approach to wildfire prevention and landscape health.

6.2.5.2. Sacred Groves in India

In India, sacred groves represent a unique form of biodiversity conservation, maintained by local communities who protect these areas due to their religious significance. These groves serve as vital reservoirs of biodiversity, housing a wide variety of flora and fauna. Their preservation not only supports ecological health but also highlights the important role of cultural beliefs in environmental conservation.

6.2.5.3. Agroforestry in Africa

Agroforestry practices in Africa exemplify sustainable agriculture through the integration of crops and trees. This traditional method enhances soil fertility, reduces erosion, and increases resilience to climate variability. By combining agricultural production with tree cultivation, communities can improve food security and promote environmental sustainability, demonstrating the value of traditional knowledge in contemporary agricultural practices.

Indigenous and local knowledge systems offer invaluable insights into environmental management, sustainability, and resilience. By recognizing and integrating these knowledge systems into broader environmental strategies, societies can benefit from a more holistic and effective approach to addressing environmental challenges. Respectful collaboration, protection of intellectual property, and equitable inclusion of traditional knowledge are essential for achieving positive outcomes and fostering sustainability (Kashif et al., 2023; Cairns, 2007; Brugnach et al., 2017).

6.3. Democratization of Environmental Knowledge

The democratization of environmental knowledge involves making information and understanding about the environment accessible to a broader range of people and communities. This process aims to empower individuals and groups by providing them with the tools, resources, and opportunities to participate in environmental decision-making and action (Forsyth, T. 2004, 2012; Anesa, 2018).

6.3.1. Definition and Objectives

The democratization of environmental knowledge refers to the process of making information about the environment accessible, understandable, and usable for everyone, regardless of their social, economic, or geographic background. This approach aims to eliminate barriers that prevent people from accessing crucial environmental information and encourages inclusive participation in managing and making decisions about environmental issues (Healy, 2009; Pereira and Laube, 2018; Moody, 2016).

The primary objectives of democratizing environmental knowledge include:

- *Inclusivity:* Ensuring that all individuals, particularly those from marginalized and underrepresented groups, have access to vital environmental information.
- *Empowerment:* Enabling people to make informed choices and take effective action concerning environmental challenges.
- *Equity:* Working to address and reduce disparities in access to environmental knowledge and resources.
- *Collaboration*: Encouraging cooperation among various stakeholders, including governments, communities, researchers, and the private sector, to share knowledge and work together towards sustainable solutions.

6.3.2. Key Strategies for Democratization

6.3.2.1. Open Access and Transparency

Open access and transparency are crucial components of democratizing environmental knowledge. One approach is through open data initiatives, which involve creating public databases and repositories that provide access to environmental data, research findings, and monitoring results. Additionally, data-sharing platforms can be established where researchers, policymakers, and the public can collaborate by sharing and accessing data from various sources (Ott and Kiteme, 2016; Guo, 2021).

Another important aspect is open-access publishing, which promotes the availability of scientific research and environmental studies to a broader audience. By ensuring that research publications are freely accessible, more individuals can engage with the findings. Furthermore, providing free access to educational materials, reports, and tools supports environmental learning and awareness among diverse communities.

6.3.2.2. Community Engagement and Participatory Approaches

Community engagement is essential for inclusive environmental management. Citizen science projects actively involve the public in scientific research, enabling individuals to contribute to data collection, monitoring, and analysis. Additionally, establishing local environmental observatories allows community members to track and report on local conditions, fostering a sense of ownership and responsibility for their environment.

Participatory decision-making processes also play a vital role in democratizing environmental knowledge. Engaging a diverse range of stakeholders, including local communities, indigenous peoples, and civil society organizations, ensures that various perspectives are considered in environmental decisions. Conducting public consultations and workshops further allows for gathering input from different groups, integrating their insights into policy and management plans.

6.3.2.3. Education and Capacity Building

Education and capacity building are key to empowering individuals and communities. Integrating environmental education into school curricula raises awareness and knowledge from an early age, while community programs such as workshops and training sessions enhance understanding of environmental issues and potential solutions.

In terms of capacity building, providing training and resources helps develop skills in areas like data analysis, environmental monitoring, and advocacy. It is also important to ensure that communities have access to the necessary tools, technologies, and support, enabling them to actively engage in effective environmental management.

6.3.2.4. Technological Solutions

Technological solutions can significantly enhance access to environmental knowledge. Digital platforms like online portals and mobile apps can provide easy access to environmental information, tools, and resources. Social media platforms can also help share environmental knowledge, engage the public, and raise awareness about pressing issues.

Moreover, leveraging crowdsourcing can gather environmental data and input from diverse contributors. Developing and promoting innovative tools and technologies further facilitates access to and utilization of environmental knowledge, making it more widely available and applicable in various contexts.

6.3.3. Challenges and Barriers

Access and equity are critical considerations in the democratization of environmental knowledge. One major challenge is the digital divide, which refers to the disparities in access to technology and internet connectivity. These disparities can significantly limit the ability of some communities to engage in the democratization of knowledge. Addressing this issue is essential to ensure that all individuals have the opportunity to participate fully in environmental discussions and actions.

Additionally, barriers related to language and literacy must be overcome to enhance the accessibility of environmental information. Variations in language and differing education levels can impede understanding, making it crucial to provide information in multiple languages and formats that cater to diverse literacy levels (O'Brien, 2013; Georgieva, 2019; Tomain, 2015). By addressing these barriers, a broader audience can engage with environmental issues.

6.3.3.1. Quality and Reliability

The quality and reliability of information shared is paramount for effective engagement. Ensuring data accuracy is essential to avoid the spread of misinformation and confusion among the public. This involves rigorous vetting processes to confirm that the information provided is factual and trustworthy.

Furthermore, addressing potential bias and representation is crucial. Information should reflect diverse perspectives and knowledge systems, ensuring that the voices of marginalized and underrepresented communities are included. This approach not only enriches the knowledge base but also fosters inclusivity in environmental discussions.

6.3.3.2. Policy and Governance

Effective policy and governance frameworks play a vital role in the democratization of environmental knowledge. It is essential to navigate regulatory barriers that may restrict access to environmental data or limit participatory approaches. Understanding and addressing these barriers can help facilitate greater access and engagement.

Moreover, securing institutional support from policymakers and organizations is necessary to advance the democratization process. By integrating democratization into broader environmental strategies, institutions can create an enabling environment that encourages participation and collaboration across various stakeholder groups. This support can also help ensure that the voices of local communities are heard in decision-making processes.

6.3.4. Case Studies and Examples

6.3.4.1. Global Forest Watch

Global Forest Watch is a comprehensive platform that offers open access to satellite data related to deforestation, forest degradation, and land use. This innovative tool enables users to monitor and analyze changes in forest conditions in real time. By providing timely information, Global Forest Watch empowers governments, NGOs, and local communities to take informed actions in forest conservation and management.

6.3.4.2. iNaturalist

iNaturalist is a citizen science app designed to engage users in recording and sharing their observations of biodiversity. By allowing individuals to contribute their findings, iNaturalist not only fosters public participation in scientific research but also helps build global biodiversity databases. This platform creates a community of nature enthusiasts and scientists working together to document and protect the planet's diverse ecosystems.

6.3.4.3. Local Environmental Action Groups

Local environmental action groups are grassroots organizations focused on addressing environmental issues within their communities. These initiatives often leverage local knowledge and collaborative approaches to tackle challenges such as pollution, conservation efforts, and climate adaptation. By involving community members in decision-making and action, these groups play a vital role in promoting sustainable practices and fostering resilience at the local level.

The democratization of environmental knowledge is essential for creating a more inclusive, equitable, and effective approach to environmental management and sustainability. By making information more accessible, involving diverse stakeholders, and leveraging technological solutions, societies can empower individuals and communities to play a meaningful role in addressing environmental challenges (Lim and Tang, 2002; Forsyth, 2004; Fischer, 1993). Overcoming barriers and addressing challenges related to access, quality, and policy will be crucial for achieving successful outcomes and ensuring that the benefits of democratized knowledge are widely shared.

6.4. Science Policy Interfaces in Environmental Governance

The science-policy interface is a crucial element in environmental governance, bridging scientific research and policy-making processes. It involves exchanging knowledge, information, and insights between scientists, policymakers, and other stakeholders to inform and guide effective environmental policies and practices (Koetz et al., 2012; Sarkki et al., 2020; Haas, 2017).

6.4.1. Definition and Importance

The science-policy interface refers to the mechanism that facilitates the translation of scientific knowledge into policy and decision-making processes. This interface involves active collaboration between researchers, policymakers, and other relevant stakeholders to ensure that the best available scientific evidence informs environmental policies effectively (Never, 2012; Saner, 2013; Monkelbaan, 2019).

The science-policy interface is crucial for promoting evidence-based policy. By grounding environmental policies in scientific research and data, this approach enhances their effectiveness and relevance. Additionally, it supports informed decision-making, allowing policymakers to utilize reliable information when addressing issues related to environmental management, regulation, and conservation. Furthermore, this interface plays a vital role in problem-solving, as it helps tackle complex environmental challenges by integrating diverse sources of knowledge and expertise.

6.4.2. Key Components of Science-Policy Interfaces

Knowledge production and dissemination involves two main components: research and data collection and the communication of findings. Scientific research is essential for conducting studies and gathering data on various environmental issues, trends, and impacts. Ongoing environmental monitoring and assessment are also crucial, as they inform policy and management strategies by providing timely updates on environmental changes (Glaser and Glaeser, 2014, 2023).

Effective communication of research findings is vital for bridging the gap between science and policy. This involves publishing research results in scientific journals, reports, and policy briefs, making them accessible to policymakers and the general public. Additionally, utilizing media, conferences, and outreach activities can further disseminate scientific knowledge and highlight its implications for policy decisions.

6.4.2.1. Policy Development and Implementation

Policy development and implementation consist of two key stages: policy formulation and policy implementation. Integrating scientific findings into the policy development process is crucial in the formulation stage to ensure that environmental issues are effectively addressed. Engaging with scientists, experts, and stakeholders for input and feedback during this stage can lead to more informed and robust policies.

Once policies are formulated, the next step is their implementation. This involves developing and executing action plans based on scientific recommendations to tackle specific environmental challenges. Continuous monitoring and evaluation are also necessary to assess the effectiveness of these policies, allowing for adjustments based on new scientific information and data.

6.4.2.2. Stakeholder Engagement

Stakeholder engagement is a critical aspect of the science-policy interface, encompassing various mechanisms and collaborative platforms. One effective method is the establishment of advisory committees or panels comprised of scientists, experts, and stakeholders who can provide valuable advice and recommendations on policy issues. Additionally, public consultation is essential for involving diverse perspectives and knowledge in the policy process, ensuring that various interests and voices are considered.

Collaborative platforms, such as partnerships between research institutions, government agencies, NGOs, and other stakeholders, facilitate knowledge exchange and collaborative problem-solving. Moreover, creating networks and forums for ongoing dialogue and interaction between scientists and policymakers can foster stronger relationships and enhance the effectiveness of environmental policies (Glaser and Glaeser, 2014, 2023).

6.4.3. Models and Approaches

Science-policy platforms play a crucial role in bridging the gap between scientific knowledge and policymaking. These platforms can be categorized into intergovernmental panels and national or regional platforms (Mace et al., 2018; Biggs et al., 2023; Capano and Jarvis, 2020).

Intergovernmental Panels include organizations like the Intergovernmental Panel on Climate Change (IPCC) and the Intergovernmental Science-Policy Platform on Biodiversity and Ecosystem Services (IPBES). The IPCC provides policy-relevant scientific assessments on climate change, which are vital for informing international negotiations and actions. Similarly, IPBES delivers assessments and knowledge on biodiversity and ecosystem services to support effective policy and decision-making (Koetz et al., 2012; Sarkki et al., 2020; Haas, 2017).

On national and regional levels, environmental protection agencies play a significant role in integrating scientific research into environmental regulations and policies. Additionally, regional initiatives focus on addressing specific environmental issues by incorporating scientific knowledge into local policy and management practices.

6.4.3.1. Knowledge Brokering

Knowledge brokering refers to the processes that facilitate the translation of scientific knowledge into practical policy solutions. This includes the work of science advisors and the establishment of knowledge networks.

Science Advisors are individuals or teams within government or policy organizations who provide scientific expertise and guidance to inform policy decisions. They often produce research syntheses that summarize and synthesize scientific findings into actionable insights and recommendations for policymakers.

Knowledge Networks consist of expert networks that connect researchers and policymakers, providing advice and conducting assessments on environmental issues. Collaborative research initiatives further strengthen this connection by fostering joint research projects between scientists and policymakers, specifically aimed at addressing pressing environmental challenges.

6.4.4. Challenges and Barriers

Communication gaps often hinder the effective integration of scientific knowledge into policy decisions. One of the primary challenges is the complexity of science, where conveying intricate scientific concepts and findings in a manner that is both understandable and actionable for policymakers proves difficult. Additionally, the issue of information overload complicates the situation further. Policymakers must manage and interpret vast amounts of scientific data and research findings, which can overwhelm their ability to make informed decisions (Sarkki et al., 2020; Weichselgartner and Kasperson, 2010; Koetz et al., 2012).

6.4.4.1. Institutional and Political Factors

Several institutional and political factors contribute to the challenges in the science-policy interface. Policy resistance is a significant barrier, as there may be reluctance to incorporate scientific evidence into policy due to political, economic, or ideological influences. Furthermore, fragmentation within environmental governance can exacerbate these issues. A lack of coordination and collaboration among various institutions and sectors often leads to inefficiencies and missed opportunities for integrating scientific knowledge into effective policies.

6.4.4.2. Equity and Inclusivity

Addressing equity and inclusivity is crucial in ensuring that the science-policy interface is effective and representative. This involves ensuring that all stakeholders, particularly marginalized and underserved communities, have access to environmental knowledge and opportunities to contribute to discussions. Participatory processes are essential for involving diverse stakeholders in the science-policy interface, as this approach helps to ensure that various perspectives and needs are acknowledged and considered in policy formulation and decision-making.

6.4.5. Case Studies and Examples

6.4.5.1. Climate Change Adaptation Planning

One notable case study in climate change adaptation planning involves cities and regions that have effectively utilized climate science to formulate adaptation strategies. For instance, some urban areas have integrated climate projections into their urban planning and infrastructure design processes. This proactive approach enables them to anticipate and mitigate potential climate impacts, ensuring that infrastructure is resilient to extreme weather events, rising sea levels, and other climate-related challenges.

6.4.5.2. Biodiversity Conservation

Another significant case study focuses on biodiversity conservation policies shaped by scientific assessments of species and ecosystem health. These assessments provide critical insights that inform conservation strategies, leading to the establishment of protected areas and targeted conservation programs. By relying on robust scientific data, policymakers can prioritize regions that require protection, implement effective management practices, and enhance biodiversity preservation efforts.

6.4.5.3. Air Quality Management

In the realm of air quality management, a compelling case study illustrates how the development of air quality regulations and standards has been driven by scientific research on pollutants and their health impacts. By leveraging scientific findings, governments have been able to craft regulations that effectively reduce air pollution levels. These measures have not only improved air quality but have also contributed to better public health outcomes, showcasing the vital role of science in shaping effective environmental policies.

The science-policy interface is critical in environmental governance, as it ensures that the best available scientific evidence informs policies. Integrating scientific knowledge into policy-making processes enhances the ability to address environmental challenges, supports informed decision-making, and promotes sustainable outcomes. Overcoming communication, institutional factors, and equity challenges will be essential for improving the effectiveness and inclusivity of the science-policy interface (Sarkki et al., 2020; Koetz et al., 2012; Saarela, 2020). By fostering collaboration and engagement between scientists, policymakers, and stakeholders, societies can better navigate complex environmental issues and develop effective solutions.

6.5. The Role of Big Data and AI in Environmental Monitoring

Big Data and Artificial Intelligence (AI) have become transformative tools in environmental monitoring, offering advanced capabilities for collecting, analyzing, and interpreting vast amounts of environmental data. These technologies enhance our ability to understand environmental changes, manage natural resources, and address complex environmental challenges (Himeur et al., 2022; Vagin et al., 2022; Bibri et al., 2023).

6.5.1. Big Data in Environmental Monitoring

Big Data refers to the vast amounts of data generated at high speed and from diverse sources, including sensors, satellites, social media, and various digital platforms. Its characteristics include large scale, rapid generation, and a wide range of formats, encompassing both structured and unstructured data. The complexity and volume of Big Data necessitate advanced tools and techniques for effective processing and analysis, allowing for insights that would otherwise remain hidden (Misra et al., 2020; Benzidia et al., 2021; Himeur et al., 2023).

6.5.1.1. Sources of Big Data

One significant source of Big Data is remote sensing, which includes data collected from satellites and drones. Satellites provide high-resolution images and valuable information regarding land use, vegetation, and weather patterns. Drones enhance this data collection by capturing detailed aerial imagery of specific areas, particularly in remote or difficult-to-access locations.

Another important source of Big Data comes from sensors and Internet of Things (IoT) devices. Environmental sensors monitor critical parameters such as air and water quality, temperature, humidity, and pollution levels. Smart meters contribute to this data landscape by collecting information on energy and water consumption, thereby aiding in effective resource management.

Social media and crowdsourced data also play a vital role in generating Big Data. User-generated content from social media platforms and crowdsourcing applications gathers information about environmental conditions and events, providing real-time insights and community engagement.

6.5.1.2. Applications of Big Data

The applications of Big Data are vast and impactful, particularly in environmental monitoring. Real-time tracking systems enable continuous observation of environmental parameters, such as air and water quality, facilitating timely detection of changes and anomalies. Additionally, trend analysis of long-term data helps identify patterns and assess the effects of environmental changes, allowing for proactive management.

In disaster management, Big Data is invaluable for creating early warning systems that utilize information from multiple sources to predict and mitigate the impacts of natural disasters, such as floods, hurricanes, and wildfires. Furthermore, it enhances disaster response coordination by providing real-time information and situational awareness to emergency responders.

Big Data also supports biodiversity and ecosystem management through applications like species tracking and habitat analysis. Monitoring the distribution and behavior of species aids conservation efforts and protects biodiversity, while analyzing habitat conditions and changes informs effective ecosystem management and restoration initiatives.

6.5.2. AI in Environmental Monitoring

Artificial Intelligence (AI) encompasses the development and utilization of advanced computational techniques such as machine learning and deep learning to analyze, interpret, and understand data. It involves the creation of systems that can recognize patterns, make predictions, and even learn from experience without human intervention. Various techniques underpin AI, including algorithms and models like neural networks, which mimic the workings of the human brain, decision trees that help in making informed choices based on data attributes, and natural language processing (NLP) that allows machines to understand and generate human language. These techniques form the backbone of AI applications, enabling machines to process vast amounts of data efficiently and accurately (Himeur et al., 2022; Zhang et al., 2021; Rodriguez and Costa, 2024).

6.5.2.1. Applications of AI

AI has numerous applications across various domains, significantly enhancing capabilities in data analysis and pattern recognition. One prominent use of AI is in image recognition, where algorithms analyze satellite and drone imagery to identify different land cover types, detect changes in landscapes, and assess environmental conditions effectively. Another critical application is anomaly detection, which involves identifying unusual patterns or outliers in environmental data, such as spikes in pollution levels or alterations in habitats. By employing these AI-driven techniques, researchers and environmental managers can gain valuable insights that inform their actions and policies.

Predictive modeling is another area where AI demonstrates its potential, particularly in forecasting climate conditions and trends. By utilizing historical data along with machine learning models, AI can provide accurate climate predictions, helping communities prepare for future environmental changes. Additionally, AI can develop species distribution models that predict how species might respond to shifts in environmental variables and climate change scenarios. Such predictive capabilities are essential for conservation efforts and biodiversity management.

AI also plays a vital role in decision support systems, where it aids resource management and policy formulation. By providing data-driven insights and recommendations, AI enhances decision-making in critical areas such as water allocation, land use planning, and conservation strategies. Furthermore, it assists policymakers in developing and evaluating environmental policies by modeling potential outcomes and impacts, ensuring that decisions are informed by robust evidence.

6.5.2.2. Integration with Big Data

The integration of AI with Big Data significantly enhances data processing capabilities. AI technologies enable real-time analytics, allowing for the processing of large volumes of data from diverse sources instantly. This capability provides immediate insights that facilitate swift responses to emerging environmental issues. Additionally, AI excels in data fusion, combining information from different sources and formats to create comprehensive environmental assessments and models. This holistic approach leads to a deeper understanding of environmental dynamics and challenges.

Moreover, AI contributes to automation and efficiency in environmental monitoring and management. Automated reporting systems can generate reports and visualizations based on data analysis, minimizing the need for manual processing and interpretation. This not only saves time but also enhances accuracy in data reporting. Operational efficiency is further optimized through the automation of routine tasks and data analysis processes, allowing human experts to focus on more complex issues. Together, these advancements in AI and Big Data create a powerful synergy that improves environmental management and decision-making.

6.5.3. Challenges and Considerations

Ensuring data quality and accuracy is paramount in the field of environmental management, particularly when leveraging AI and big data. Data integrity involves maintaining the accuracy and reliability of the information collected from various sources, which is essential for making informed decisions. Without high-quality data, analyses may yield misleading results, ultimately impacting the effectiveness of environmental policies and interventions. Additionally, it is crucial to address potential biases and errors inherent in AI algorithms and the data sources themselves. These biases can stem from unrepresentative data sets or flawed assumptions within the algorithms, potentially leading to skewed analysis and misguided decision-making (Binetti et al., 2024; Ramadhani, 2024; Karthikeyan et al., 2022). By implementing robust data validation and correction processes, stakeholders can enhance the reliability of their findings and ensure that the insights generated are truly reflective of the environmental conditions being studied.

6.5.3.1. Privacy and Security

The protection of sensitive environmental and personal data is another critical aspect of employing AI and big data in environmental management. Data privacy must be prioritized to prevent unauthorized access and misuse, which could compromise not only individual privacy but also public trust in data initiatives. This involves establishing clear protocols for data handling, storage, and sharing that adhere to relevant legal and ethical standards. Cybersecurity is equally vital, as environmental data and systems face an increasing number of cyber threats and attacks. Implementing comprehensive cybersecurity measures helps safeguard against potential breaches, ensuring the integrity of both the data and the systems used to analyze and manage it. By prioritizing privacy and security, organizations can create a more resilient framework for environmental data management.

6.5.3.2. Interoperability and Integration

For AI and big data tools to be effective in environmental monitoring and management, interoperability and integration are essential. Data standardization plays a crucial role in this process by developing consistent formats and protocols that facilitate the integration and analysis of data across different platforms and systems. Standardized data enables various stakeholders to share information more easily and collaborate effectively, leading to richer analyses and more comprehensive insights. Furthermore, ensuring seamless integration of AI and big data tools with existing environmental monitoring and management systems is vital. This requires thoughtful planning and execution to create compatible systems that can work together, allowing for the efficient flow of information and maximizing the utility of available data. By focusing on interoperability and integration, organizations can enhance their overall capacity for effective environmental management and decision-making.

6.5.4. Case Studies and Examples

6.5.4.1. Global Forest Watch

Global Forest Watch is an innovative application that harnesses satellite imagery and artificial intelligence (AI) to monitor deforestation and forest degradation in real-time. This platform provides crucial alerts and data to support conservation efforts and inform policy interventions aimed at protecting forest ecosystems. By analyzing high-resolution satellite images, Global Forest Watch can detect changes in forest cover, identify areas at risk of deforestation, and provide actionable insights for stakeholders involved in forest management. This timely information empowers governments, NGOs, and local communities to take proactive measures to address deforestation and promote sustainable land use practices.

6.5.4.2. IBM's Green Horizons Project

IBM's Green Horizons Project is a cutting-edge initiative that leverages AI and big data analytics to address urban air quality challenges. By analyzing vast amounts of air quality data, the project can predict pollution levels and develop effective strategies for improving air quality in cities. This application not only enhances our understanding of pollution sources and trends but also supports policymakers in implementing data-driven solutions to mitigate air quality issues. Through predictive modeling and real-time monitoring, IBM's Green Horizons Project aims to create healthier urban environments and foster sustainable practices that protect public health.

6.5.4.3. eBird

eBird is a pioneering application that utilizes crowdsourced bird observation data to track migration patterns and monitor biodiversity. By analyzing the contributions of birdwatchers and citizen scientists, eBird employs AI algorithms to identify trends in bird populations and their movements across various habitats. This data is invaluable for informing conservation strategies and understanding the impacts of environmental changes on avian species. By engaging the public in scientific research, eBird fosters a greater appreciation for biodiversity and empowers individuals to contribute to conservation efforts. Through this collaborative approach, eBird enhances our knowledge of bird populations and supports initiatives aimed at protecting and preserving wildlife.

Big Data and AI are revolutionizing environmental monitoring by enhancing data collection, analysis, and interpretation capabilities. These technologies provide valuable insights into environmental conditions, support effective management and decision-making, and address complex environmental challenges (Jumaev, 2024; Hsu et al., 2022; Adeleke et al., 2017). Addressing data quality, privacy, and integration challenges will be crucial for maximizing the benefits of these technologies and ensuring their responsible use in environmental governance.

Chapter 7

Environmentalism and Social Transformation

Environmentalism and social transformation are deeply interconnected, with each influencing and reshaping the other. Environmentalism seeks to address ecological issues and promote sustainable practices, while social transformation involves fundamental changes in societal structures, values, and behaviors. The intersection of these fields highlights how environmental issues can drive social change and how social factors can influence environmental policies and actions (Epstein, 1995; Coates, 2003; Van Rensburg, 1994).

7.1. The Interconnection Between Environmentalism and Social Transformation

The relationship between environmental challenges and social change is characterized by a dynamic interplay where each influences the other in significant ways. Environmental crises such as climate change, pollution, and biodiversity loss often serve as catalysts for substantial shifts in societal attitudes, behaviors, and policies. As communities grapple with the tangible impacts of these crises, there is typically a corresponding rise in awareness and activism. Individuals and groups may mobilize to advocate for systemic changes aimed at addressing environmental degradation and promoting sustainable practices (Peet and Watts, 2004; Fischer-Kowalski and Haberl, 2007; Clapp and Dauvergne, 2011). This heightened sense of urgency can drive demands for more robust environmental policies, enhanced regulations, and greater accountability from governments and corporations.

Conversely, social changes can significantly impact environmentalism. Social movements, shifting public opinion, and evolving cultural values often shape the direction of environmental policies and practices. For instance, increased awareness of social justice issues may lead to a greater emphasis on equity within environmental initiatives, ensuring that marginalized communities are considered in decision-making processes. As society transforms, there can be a growing demand for sustainable practices that prioritize not only ecological health but also social well-being. This reciprocal relationship highlights the importance of understanding how environmental challenges and social changes coexist and influence each other, ultimately shaping a more sustainable and equitable future for all.

7.1.1. Key Areas of Social Transformation Driven by Environmentalism

The increasing awareness of environmental issues has prompted significant cultural and behavioral shifts across societies. One notable change is the rise of sustainable lifestyles, which reflects a growing commitment to eco-friendly practices. As consumers become more conscious of their environmental footprint, they are altering their consumption patterns to prioritize eco-friendly products. This shift includes a preference for items that are sustainably sourced, efforts to reduce waste, and a general focus on sustainable living principles. Individuals are making more deliberate lifestyle choices, such as reducing their energy consumption, supporting local and organic agriculture, and minimizing the use of single-use plastics (Epstein, 1995; McGrail, 2011; Kassas, 1997; Bhandari, 2019). These changes not only reflect personal values but also contribute to broader societal movements toward sustainability.

In tandem with these lifestyle changes, there has been a significant increase in environmental awareness and education. Environmental movements and campaigns have played a pivotal role in raising public awareness about pressing ecological issues, leading to a greater understanding of the interconnectedness of human activities and environmental health. Educational institutions, including schools, universities, and community organizations, are integrating environmental education into their curricula, fostering awareness and encouraging action from an early age. By equipping individuals with the knowledge and skills needed to engage with environmental challenges, these educational initiatives are nurturing a generation that is better prepared to address the complexities of sustainability.

7.1.1.1. Policy and Institutional Changes

The societal push for environmental protection has also catalyzed notable policy and institutional changes. A crucial aspect of this transformation is the development and implementation of various environmental laws and regulations. These legislative measures aim to reduce pollution, conserve natural resources, and mitigate the effects of climate change. International agreements, such as the Paris Agreement and the Convention on Biological Diversity, exemplify the collective commitments made by countries to address environmental challenges and promote sustainable development on a global scale. Such frameworks facilitate collaboration among nations and underscore the importance of shared responsibility in protecting the planet.

Institutional reforms are another vital component of the evolving environmental landscape. Corporations are increasingly adopting corporate social responsibility (CSR) practices, integrating sustainability into their operations and supply chains. This shift not only enhances corporate accountability but also aligns business practices with broader societal goals. Additionally, the growing recognition of environmental issues has led to the establishment of dedicated government agencies and departments focused on environmental protection and sustainable development. These institutions play a crucial role in shaping and enforcing environmental policies, ensuring that sustainability is prioritized in decision-making processes.

7.1.1.2. Social Equity and Justice

Amid these transformations, the principles of social equity and justice have gained prominence within the environmental movement. Environmentalism has underscored the necessity of equitable distribution of environmental benefits and burdens. This focus on environmental justice addresses critical issues such as pollution exposure and access to green spaces, particularly for marginalized communities. Advocacy for these communities is essential, as they often bear the brunt of environmental degradation and climate impacts. Environmental justice movements work to amplify the voices of those disproportionately affected, ensuring that their rights and needs are recognized and addressed in policy discussions.

Community empowerment is another vital aspect of this evolving landscape. Local initiatives have emerged as communities take proactive measures to address environmental issues that directly affect them. Examples include organizing clean-up drives, establishing community gardens, and advocating for sustainable practices within their neighborhoods. Furthermore, social transformation often involves enhancing community participation in environmental decision-making. By incorporating local voices and needs into policy and planning processes, communities can more effectively advocate for solutions that reflect their unique contexts and challenges. This participatory governance approach fosters a sense of ownership and responsibility, empowering individuals to contribute actively to environmental stewardship and sustainable development.

7.1.2. The Role of Social Movements in Environmentalism

Grassroots movements play a crucial role in the environmental landscape by fostering activism and mobilizing communities for environmental advocacy. These movements often focus on specific issues such as climate change, conservation, and pollution, highlighting the urgency of these challenges and the need for concerted action. Activists within these movements engage in a range of activities, including organizing campaigns, protests, and awareness-raising initiatives aimed at mobilizing public support and influencing policy changes (Yearley, 2013; Haluza-DeLay, 2008; Coglianese, 2001). By harnessing the collective power of individuals who share a common goal, grassroots movements can effectively challenge established practices and advocate for more sustainable and equitable policies.

In addition to advocacy, grassroots organizations are instrumental in implementing community-based solutions. They often initiate local projects and initiatives designed to tackle specific environmental challenges while promoting sustainable practices within their communities. These projects may include reforestation efforts, waste reduction programs, and community gardens that encourage local food production. Beyond project implementation, grassroots movements also prioritize capacity building, empowering communities to take informed action on environmental issues. By providing education, resources, and support, these movements enhance local knowledge and skills, enabling individuals and groups to become effective advocates for sustainability in their own right.

7.1.2.1. Global Movements

Global movements have emerged as powerful forces in the fight against environmental degradation, particularly through youth activism and international solidarity. One of the most prominent examples is the Fridays for Future movement, which has mobilized millions of young people around the world to demand urgent action on climate change. Fueled by a shared commitment to securing a sustainable future, these youth movements emphasize the critical importance of intergenerational justice and the need for immediate policy changes to combat climate change. Their activism often manifests in organized strikes, public demonstrations, and other forms of civic engagement, drawing attention to the urgency of environmental issues and inspiring others to join the cause.

Moreover, global environmental movements foster international solidarity and cooperation to address transboundary environmental issues.

These movements recognize that environmental challenges often cross-national borders, necessitating collective action and collaboration among nations. By bringing together diverse stakeholders, including governments, NGOs, and civil society, global movements work to promote shared goals and encourage the development of international agreements that prioritize environmental protection.

Non-governmental organizations (NGOs) and advocacy groups also play a critical role in shaping environmental policy and driving social and political change on a global scale. These organizations engage in a wide range of activities, including conducting research, raising public awareness, and lobbying for legislative changes. By leveraging their expertise and networks, NGOs can influence policymakers and ensure that environmental considerations are integrated into decision-making processes. Additionally, they often serve as vital resources for grassroots movements, providing support, training, and funding to help amplify local voices and initiatives. Through these combined efforts, grassroots and global movements work synergistically to promote a more sustainable and just world.

7.1.3. Challenges and Opportunities in Integrating Environmentalism with Social Transformation

Despite the growing momentum toward environmental sustainability, several significant challenges impede progress. One major hurdle is resistance to change, which often arises from entrenched political and economic interests. Established entities may resist alterations that threaten their profitability or power, resulting in conflicts that hinder the implementation of effective environmental policies. For instance, industries that rely heavily on fossil fuels may lobby against regulations aimed at reducing carbon emissions, thereby obstructing necessary reforms. Additionally, deeply ingrained cultural norms and values can pose barriers to adopting new environmental practices and perspectives (Kuenkel et al., 2019; Al-Shetwi, 2022; Akhtar, 2011). Societies with long-standing traditions or beliefs may struggle to reconcile these with contemporary environmental demands, leading to pushback against innovative solutions.

Another challenge lies in inequality and exclusion within the environmental decision-making process. Certain social groups, particularly marginalized communities, may find themselves excluded from discussions and decisions that directly affect their lives and environments. This exclusion can result in inequitable outcomes that exacerbate existing inequalities, leaving vulnerable populations with limited resources to address environmental issues. Moreover, disparities in access to resources and opportunities can restrict the ability of specific communities to participate in and benefit from environmental initiatives. Without meaningful inclusion and representation, efforts to create sustainable solutions risk failing to address the diverse needs and priorities of all stakeholders.

7.1.3.1. Opportunities

Amid these challenges, there are also significant opportunities for advancing environmental sustainability. One promising avenue is through innovation and collaboration. Cross-sector collaboration among environmental organizations, businesses, governments, and communities can lead to innovative solutions and effective strategies for tackling environmental and social challenges. By pooling resources, expertise, and perspectives, stakeholders can develop comprehensive approaches that address the multifaceted nature of environmental issues. Additionally, leveraging technology and data can enhance the effectiveness of environmental initiatives. Advanced analytics, remote sensing, and AI can provide valuable insights for informed decision-making and monitoring, ensuring that interventions are data-driven and targeted.

Furthermore, building resilient communities presents another vital opportunity. Integrating environmental and social goals can foster resilient and sustainable communities that are better equipped to adapt to environmental changes and challenges. By prioritizing sustainable development, communities can enhance their capacity to withstand environmental shocks while promoting social equity. Empowerment and inclusion are also critical components of this process. Promoting inclusive and participatory approaches ensures that all community members have a voice in environmental decision-making, thereby enhancing social cohesion. By distributing environmental benefits equitably, these efforts can not only address environmental challenges but also contribute to a more just and resilient society. Through collaboration, innovation, and inclusive practices, stakeholders can navigate the complexities of environmental sustainability and work toward a more sustainable future for all.

7.1.4. Case Studies and Examples

7.1.4.1. Community-Led Conservation Initiatives

A noteworthy example of community-led conservation initiatives is the Maasai Mara Community Conservancies in Kenya. In this region, local communities have taken the initiative to establish conservancies dedicated to protecting wildlife and managing natural resources sustainably. This collaborative approach has led to significant improvements in conservation outcomes, including the protection of endangered species and the restoration of habitats. Furthermore, the conservancies have generated tangible benefits for the communities involved, such as economic opportunities through eco-tourism and enhanced livelihoods. By integrating traditional knowledge and practices with modern conservation strategies, the Maasai Mara Community Conservancies serve as a model for how local stewardship can effectively contribute to biodiversity preservation and sustainable resource management.

7.1.4.2. Urban Sustainability Movements

Urban sustainability movements have gained traction in cities worldwide, with Copenhagen and Portland standing out as exemplary cases. These cities have embraced green urban planning principles, focusing on sustainable transportation, energy efficiency, and community engagement to create livable and resilient urban environments. Copenhagen, for instance, has invested heavily in cycling infrastructure and public transportation, significantly reducing its carbon footprint and promoting a culture of sustainability among residents. Similarly, Portland has implemented innovative policies aimed at reducing waste, enhancing green spaces, and fostering local food systems. The collective efforts of local governments, businesses, and communities in these cities have resulted in not only reduced environmental impacts but also improved quality of life for residents, highlighting the importance of collaborative approaches in urban sustainability.

7.1.4.3. Global Climate Strikes

The Fridays for Future movement exemplify the power of global climate strikes in mobilizing public awareness and action around climate change. Spearheaded by young activists like Greta Thunberg, this grassroots movement has inspired millions of people worldwide to participate in strikes, demanding urgent action from governments to address the climate crisis. The movement has successfully raised awareness about the urgency of climate action and the need for systemic changes to mitigate climate impacts. As a result, it has influenced policy discussions, encouraging governments and institutions to prioritize climate action in their agendas. By uniting diverse voices and fostering a sense of urgency, the Fridays for Future movement illustrates the potential of youth-led activism to reshape public perceptions and drive meaningful change in environmental policies on a global scale.

Environmentalism and social transformation are interlinked, with environmental issues driving social change and social factors shaping environmental policies and actions. By addressing the challenges and leveraging the opportunities at this intersection, societies can foster more sustainable, equitable, and resilient communities (Lamperti et al., 2019; Harrison et al., 2019; Handelman and Brynen, 2019). Integrating environmental goals with social transformation efforts can lead to holistic solutions that address sustainability's ecological and social dimensions.

7.2. Environmental Justice and Equity in Transformation

Environmental justice and equity are critical dimensions of the broader environmental transformation agenda. They emphasize the fair distribution of environmental benefits and burdens and seek to address the disparities and injustices arising from environmental policies and practices (Amorim-Maia et al., 2022; Bennett et al., 2019; Agyeman et al., 2002).

7.2.1. Understanding Environmental Justice

Environmental justice is a principle that asserts the right of all individuals, irrespective of race, ethnicity, income, or socio-economic status, to equal access to a healthy environment. This principle emphasizes that everyone should be protected from environmental harm and advocates for rectifying the disproportionate impacts of environmental degradation that often fall on marginalized communities. Environmental justice seeks to create equitable conditions where all people can thrive without facing the adverse effects of pollution, toxic waste, or climate change (Menton et al., 2020; Reckien et al., 2018; Warriner et al., 2001).

7.2.1.1. Key Principles of Environmental Justice

7.2.1.1.1. Equitable Distribution of Environmental Benefits and Burdens

One of the core tenets of environmental justice is the equitable distribution of both environmental benefits and burdens. This means ensuring that all communities have access to essential environmental amenities such as clean air, safe drinking water, and green spaces. At the same time, it involves addressing the disproportionate exposure that marginalized communities face regarding environmental hazards, including pollution, toxic waste, and the impacts of climate change. By striving for fairness in how these benefits and burdens are allocated, environmental justice seeks to protect vulnerable populations from systemic inequities.

7.2.1.1.2. Meaningful Participation

Meaningful participation is another fundamental principle of environmental justice, advocating for the involvement of affected communities in environmental decision-making processes. This principle asserts that the voices of those who are most impacted by environmental policies should be heard and considered. Additionally, it emphasizes the importance of providing these communities with access to relevant information about environmental risks, regulations, and policies that affect their lives. By fostering an inclusive dialogue, environmental justice empowers communities to advocate for their needs and interests effectively.

7.2.1.1.3. Redress and Accountability

The principle of redress and accountability focuses on providing remedies for communities adversely affected by environmental harms, including health issues and loss of livelihoods. It emphasizes the need for compensation and support for those who have suffered due to environmental injustices. Furthermore, accountability entails holding responsible parties—corporations, government agencies, or other entities—responsible for their actions and ensuring that policies are implemented fairly and effectively. This principle is crucial for building trust and restoring justice in communities historically facing environmental neglect and harm.

7.2.2. The Role of Equity in Environmental Transformation

Equity refers to the fair and just distribution of resources, opportunities, and benefits, with a focus on the diverse needs and circumstances of different individuals and groups. This concept recognizes that treating everyone equally (equality) does not necessarily result in fair outcomes, as it often overlooks existing disparities and systemic inequalities. Instead, equity emphasizes the importance of addressing these disparities by providing support and resources based on individual and community needs, thus promoting justice and inclusivity in societal structures (Magliocca, 2023; Amorim-Maia et al., 2022; McDermott et al., 2023).

7.2.2.1. Key Aspects of Equity

7.2.2.1.1. Distributive Equity

Distributive equity involves the fair allocation of environmental resources and benefits, particularly to disadvantaged or marginalized groups. This aspect of equity ensures that all individuals have access to essential environmental amenities, such as clean air and water, as well as opportunities related to environmental participation, education, and employment. By prioritizing equitable resource distribution, distributive equity aims to correct historical imbalances and foster an environment where everyone can thrive.

7.2.2.1.2. Procedural Equity

Procedural equity focuses on the processes through which decisions are made regarding environmental policies and resource allocation. It emphasizes the importance of inclusive decision-making, ensuring that the perspectives and needs of all stakeholders—especially those most affected by environmental issues—are considered. Transparency in decision-making is also crucial; stakeholders should be informed about how decisions are made and how resources are allocated. By fostering inclusive and transparent processes, procedural equity builds trust and empowers communities to engage meaningfully in environmental governance.

7.2.2.1.3. Restorative Equity

Restorative equity seeks to address past inequities and injustices through targeted interventions and policies. This aspect emphasizes the need to rectify historical wrongs, particularly those that have disproportionately affected marginalized communities. Restorative equity also involves building resilience by enhancing the capacity of vulnerable communities to adapt to environmental changes and challenges. Through these efforts, restorative equity aims to create a more just and sustainable future by ensuring that past injustices are acknowledged and addressed, paving the way for a more equitable distribution of resources and opportunities moving forward.

7.2.3. Challenges and Barriers to Achieving Environmental Justice and Equity

Systemic inequities are deeply rooted in historical discrimination and long-standing inequalities that perpetuate environmental injustices. Historical patterns of discrimination—based on race, ethnicity, and socio-economic status—have created barriers that limit access to essential resources and opportunities for marginalized communities. This historical context shapes the landscape of environmental justice today, leading to disparities in health, well-being, and environmental quality. Economic disparities further exacerbate these inequities, as communities with fewer financial resources often find themselves more vulnerable to environmental harm. They may live in areas with higher pollution levels or limited access to clean water and green spaces, which further compounds their challenges and restricts their ability to thrive (Ulibarri et al., 2022; Buchanan and Wozniak-Brown, 2023; Agyeman et al., 2002).

7.2.3.1. Power Dynamics

Power dynamics play a crucial role in shaping environmental outcomes, with disparities in political influence leading to unequal representation in decision-making processes. Marginalized communities often lack the political clout necessary to advocate effectively for their needs and concerns, which can result in policies that overlook or actively harm their interests. Additionally, powerful corporate interests frequently prioritize profit over the principles of environmental justice. This can manifest in legislation and regulatory decisions that disproportionately benefit specific groups—often those with financial means—while exacerbating the struggles of disadvantaged populations. The interplay between political influence and corporate interests creates a landscape where equity in environmental governance is difficult to achieve.

7.2.3.2. Lack of Access and Representation

The lack of access and representation for marginalized communities further compounds the challenges they face in advocating for environmental justice. These communities are often underrepresented in environmental governance and decision-making bodies, leading to policies that do not reflect their needs or priorities. Additionally, limited access to information and resources hinders their ability to participate effectively in environmental advocacy. Without sufficient information about environmental risks and available resources, these communities may struggle to assert their rights or mobilize support for their causes. The combination of underrepresentation and information gaps creates significant barriers to achieving equitable outcomes in environmental policy and practice, perpetuating a cycle of disadvantages and exclusion.

7.2.4. Strategies for Promoting Environmental Justice and Equity

7.2.4.1. Policies, Legal Frameworks and Environmental Regulations

A robust policy and legal framework are essential for advancing environmental justice. Developing equitable standards involves creating and enforcing environmental regulations that protect all communities, particularly those disproportionately affected by environmental hazards. This means establishing guidelines that prioritize the health and safety of marginalized groups, ensuring that they have equal access to clean air, water, and green spaces. Additionally, compliance and enforcement mechanisms must be effective, holding violators accountable for failing to meet these standards. This accountability not only deters future violations but also builds trust within communities that their well-being is prioritized in regulatory processes (Gracia and Koh, 2011; Bullard, 2007; Bell, 2014).

7.2.4.2. Legal Protections

Incorporating environmental justice principles into legal frameworks is vital for safeguarding vulnerable communities. Rights-based approaches should be integrated into human rights agreements and national legislation to ensure that the right to a healthy environment is recognized and protected. Access to justice is also a critical component; communities must have mechanisms in place to seek redress for environmental harm they experience. This could include legal aid services, community advocacy groups, and streamlined processes for filing complaints, enabling affected individuals and communities to hold responsible parties accountable.

7.2.4.3. Community Engagement and Empowerment

Empowering communities is crucial for effective environmental justice advocacy. Capacity building involves offering education and training programs that equip individuals with the knowledge and skills needed to engage in environmental decision-making. This training can cover various topics, such as understanding environmental regulations, advocating for policy changes, and conducting community assessments. Additionally, resource support—both financial and technical—should be provided to community organizations working on environmental justice issues. This support helps enhance their capabilities, enabling them to operate effectively and advocate for their needs.

7.2.4.4. Participatory Approaches

Engaging communities in the planning and implementation of environmental policies and projects ensures that their voices are heard and their needs are prioritized. Participatory approaches facilitate active community involvement, allowing residents to contribute to decision-making processes that affect their lives. This collaboration can lead to more equitable and effective solutions that address specific local challenges. By partnering with community organizations and stakeholders, policymakers can develop and implement solutions that promote equity and justice, fostering a sense of ownership and commitment among community members.

7.2.4.5. Data and Research

The collection and analysis of disaggregated data is essential for identifying disparities and tracking progress in addressing environmental justice issues. This involves gathering data that is broken down by race, income, location, and other relevant factors to understand how different communities are impacted by environmental policies and practices. Additionally, conducting impact assessments allows for an evaluation of the effects of proposed environmental policies and projects on diverse communities, ensuring that potential disparities are recognized and addressed before implementation.

7.2.4.5. Research and Advocacy

Evidence-based advocacy is key to promoting policies and practices that advance environmental justice and equity. Utilizing research and data helps build a compelling case for necessary changes, allowing advocates to effectively communicate the importance of environmental justice issues to policymakers and the public. Furthermore, policy analysis plays a crucial role in evaluating the potential impacts of proposed legislation, ensuring that environmental justice concerns are integrated into the policy-making process. This analysis can inform advocacy efforts and provide a foundation for pushing for changes that benefit marginalized communities.

7.2.5. Case Studies and Examples

7.2.5.1. Environmental Justice Movements

7.2.5.1.1. The Flint Water Crisis

The Flint water crisis serves as a stark example of environmental injustice, where a predominantly Black and low-income community in Flint, Michigan, was exposed to severe water contamination due to inadequate infrastructure and government negligence. The decision to switch the city's water supply to the Flint River without proper treatment led to lead leaching from aging pipes, resulting in serious health consequences for residents, including elevated lead levels in children. This crisis not only illuminated the systemic inequities in environmental governance but also galvanized local and national advocacy efforts. Activists and community leaders rallied to demand equitable investments in water infrastructure and accountability for those responsible, ultimately sparking broader conversations about the need for justice in environmental policies that disproportionately affect marginalized communities.

7.2.5.1.2. The Standing Rock Sioux Tribe

The Standing Rock Sioux Tribe's opposition to the Dakota Access Pipeline (DAPL) highlighted critical issues of indigenous rights and environmental justice. The tribe's fight centered around the protection of their water resources and sacred lands, which were threatened by the pipeline's construction. The movement gained international attention, drawing supporters from various backgrounds who joined the call for respecting indigenous sovereignty and environmental stewardship. Protesters emphasized the need for meaningful consultation with indigenous communities in decision-making processes that affect their lands and resources. The Standing Rock protests became a powerful symbol of resistance against environmental exploitation and the need for justice for marginalized groups, igniting discussions about the intersection of environmentalism and indigenous rights across the globe.

7.2.5.2. Equitable Urban Planning

7.2.5.2.1. Green Space Initiatives

Cities like Portland and New York have undertaken green space initiatives to ensure equitable access to parks and recreational areas, addressing disparities in urban environments. Recognizing that access to green spaces can significantly enhance community well-being, these initiatives aim to create inclusive environments where all residents, regardless of socioeconomic status, can enjoy nature and recreational activities. For example, the New York City Department of Parks and Recreation has focused on developing parks in underserved neighborhoods, increasing greenery in areas that have historically lacked access to outdoor spaces. These efforts contribute to improved mental and physical health, promote community cohesion, and help mitigate urban heat effects, showcasing the importance of integrating equity into urban planning practices.

7.2.5.2.2. Affordable Housing and Sustainability

The integration of sustainability into affordable housing projects illustrates a commitment to both environmental and social equity. For instance, developments that incorporate green building practices and energy-efficient technologies provide low-income communities with access to sustainable living conditions while reducing their utility costs. Projects like the construction of energy-efficient homes in various cities not only address the urgent need for affordable housing but also prioritize the environmental impacts of living spaces. By ensuring that marginalized communities benefit from sustainable housing solutions, urban planners can contribute to long-term resilience against environmental challenges, fostering inclusive growth that respects both people and the planet. These initiatives exemplify how equitable urban planning can create environments that support the health and well-being of all residents.

Environmental justice and equity are essential components of environmental transformation, ensuring that all communities are protected from environmental harms and have access to environmental benefits. Addressing systemic inequities, involving communities in decision-making, and implementing policies that promote fairness and inclusion are crucial for achieving these goals (Paben, 2011; Clark and Miles, 2021; Bell, 2014). By prioritizing environmental justice and equity, societies can foster more sustainable, resilient, and equitable environments that benefit all people.

7.3. Social Movements and Grassroots Activism

Social movements and grassroots activism are pivotal in driving environmental transformation and addressing social and environmental injustices. These collective efforts reflect the power of communities to challenge existing systems, advocate for change, and create meaningful impact from the ground up (Stammers and Eschle, 2005; Tye et al., 2018; Maiba, 2005).

7.3.1. Understanding Social Movements

Social Movements are collective efforts organized by groups of people with the aim of promoting or resisting social, political, or environmental change. These movements typically respond to perceived injustices, seeking to mobilize public opinion, influence policymakers, and reshape societal norms. They are often driven by a shared vision or set of goals among participants and may operate through a variety of methods, including protests, advocacy campaigns, lobbying, and community organizing. A defining characteristic of social movements is their grassroots nature, relying on collective action and the participation of diverse stakeholders to create momentum for change. Whether seeking reforms within the existing system or advocating for revolutionary shifts, social movements are a key vehicle for societal transformation (Martin, 2015; Buechler, 2016; Giri, 1992).

Key characteristics of social movements include collective action, shared goals, and organized efforts to achieve their objectives. While some movements may focus on incremental changes, others strive for more radical shifts in societal structures. Social movements can be long-lasting or spontaneous, localized or global, but they generally rely on cooperation, solidarity, and sustained activism to achieve their objectives.

7.3.1.1. Types of Social Movements

7.3.1.1.1. Reform Movements

Reform movements seek to enact incremental changes within existing systems or structures. Rather than challenging the entire framework of society, these movements focus on improving or refining specific policies, laws, or practices to better address injustices or inefficiencies. Reform movements often work within established institutions, advocating for change through legal, political, and institutional channels.

Example: Environmental organizations advocating for stricter pollution controls or improved waste management practices exemplify reform movements. These groups typically focus on specific issues, such as reducing carbon emissions or promoting renewable energy, without seeking to completely dismantle the current economic or political systems. Their efforts are aimed at gradual improvements within the existing framework, such as enhancing regulations, encouraging corporate responsibility, or fostering public awareness.

7.3.1.1.2. Revolutionary Movements

Revolutionary movements aim for fundamental and systemic changes to societal structures, often challenging existing power dynamics and advocating for a complete transformation of the social, economic, or political system. These movements are typically driven by a vision of an entirely new societal order and often emerge in response to deep-seated inequalities or systemic injustices that cannot be addressed through incremental reforms.

Example: Movements advocating for a complete overhaul of economic systems to address climate change and environmental degradation fall under the revolutionary category. These movements call for a radical transformation of how resources are allocated, production systems are organized, and economic power is distributed. They often push for alternative economic models, such as degrowth or eco-socialism, which seek to prioritize environmental sustainability and social equity over profit and economic growth.

7.3.1.1.3. Resistance Movements

Resistance movements arise in opposition to specific policies, projects, or practices perceived as harmful, unjust, or threatening to certain communities or ecosystems. Unlike reform movements that seek to improve existing systems, resistance movements focus on halting or preventing actions that are seen as detrimental to social or environmental well-being.

Example: Indigenous and local communities resisting land grabs or resource extraction projects that threaten their land and livelihoods are prime examples of resistance movements. These communities often engage in protests, legal battles, and international advocacy to protect their rights and prevent the destruction of their land and natural resources. Resistance movements are often rooted in defending local cultures, ecosystems, and rights against external forces, such as corporations or governments, that seek to exploit them for profit.

7.3.2. Grassroots Activism

Grassroots Activism refers to community-driven efforts focused on addressing local issues and advocating for change from the ground up. It is driven by individuals or groups working within their communities to create awareness, mobilize support, and implement practical solutions. Unlike large-scale movements led by established organizations, grassroots activism is often informal and decentralized, relying on the direct engagement of local populations. The primary aim is to give a voice to those who are directly impacted by certain issues and to empower them to lead efforts for change (Bullard and Johnson, 2000; Baumgardner and Richards, 2005; Mihaylov and Perkins, 2015).

Key characteristics of grassroots activism include local engagement, direct action, and the use of grassroots networks to build support. Activists often operate outside traditional power structures, seeking to influence decision-makers by organizing at the community level. Their efforts may address environmental, social, or political issues, and they typically focus on building long-term solutions through collective action and advocacy.

7.3.2.1. Key Components of Grassroots Activism

7.3.2.1.1. Community Organizing

Community organizing is a core component of grassroots activism, which involves building networks and mobilizing support for a shared cause. Grassroots activists often begin by forming alliances with like-minded individuals, local organizations, and community groups to strengthen their collective influence. Through meetings, events, and collaborative efforts, they foster solidarity and amplify the voices of those affected by specific issues.

- *Building Networks*: Activists create and strengthen networks by connecting with others who share common goals and concerns. These networks form the foundation of grassroots movements, helping to pool resources, expertise, and manpower to tackle the issues at hand.
- *Mobilizing Support*: Grassroots activists actively engage community members, encouraging their participation in local meetings, events, and campaigns. This helps build momentum and broad-based support for specific causes, making it easier to push for meaningful change at the local level.

7.3.2.1.2. Advocacy and Campaigning

Grassroots activism often involves advocacy and campaigning to raise awareness and influence policy. Activists work to make their issues visible, drawing attention to environmental, social, or political concerns that may otherwise go unnoticed. Through outreach efforts, they educate the public, spark conversations, and demand action from policymakers.

- *Raising Awareness*: Campaigns, media outreach, and educational efforts are essential tools for raising awareness about issues. Grassroots activists use various channels to inform the public and create a sense of urgency around their causes. These campaigns often highlight injustices or environmental problems that directly affect their communities.
- *Influencing Policy*: By engaging in advocacy, grassroots activists aim to influence policy changes at local, regional, or national levels. Whether through petitions, lobbying, or direct conversations with elected officials, they seek to shape laws and regulations to address the issues that matter most to their communities.

7.3.2.1.3. Direct Action

A defining element of grassroots activism is direct action, which includes hands-on approaches to addressing issues. This may involve organizing protests or demonstrations to raise public awareness or implementing community-based projects that directly address local concerns.

- *Protests and Demonstrations*: Grassroots activists often organize public demonstrations, marches, or protests to draw attention to pressing issues and demand action from governments or corporations. These events serve as powerful symbols of collective resistance and can be pivotal in pushing for change.
- *Community Projects*: Many grassroots initiatives involve taking action at the local level to address environmental or social issues. This could include organizing clean-up drives, starting community gardens, or running educational programs to benefit the community and promote sustainable practices directly. Through these projects, grassroots movements demonstrate their commitment to solving problems with immediate, hands-on solutions.

Grassroots activism combines community organization, advocacy, and direct action to create local, people-powered movements that have the potential to influence broader societal changes.

7.3.3. The Impact of Social Movements and Grassroots Activism

Social movements are instrumental in shaping and influencing policy and legislative change by mobilizing public opinion, advocating for specific reforms, and pressuring policymakers. These movements often act as catalysts, highlighting issues that might otherwise be overlooked or deprioritized. They create awareness of the need for regulatory measures, and through sustained advocacy, can bring about concrete changes in environmental policy and governance (Bullard and Johnson, 2000; Baumgardner and Richards, 2005; Mihaylov and Perkins, 2015).

- *Influencing Legislation*: Many landmark environmental policies have been a direct result of grassroots activism. Activists organize protests, submit petitions, and engage in lobbying efforts to push for laws that protect the environment and promote sustainability. By maintaining pressure on legislators, these movements can drive forward policies that address urgent environmental concerns, such as climate change, conservation, and pollution control.

- *Success Stories*: Examples of successful policy shifts driven by grassroots activism include the Clean Air Act in the United States, which was passed following widespread public concern over air pollution. Similarly, the establishment of national parks, marine sanctuaries, and other protected areas has often been the result of concerted efforts by environmental activists advocating for the preservation of natural spaces.

7.3.3.1. Raising Awareness and Shaping Public Opinion

A key function of social movements is raising awareness and shaping how the public perceives critical environmental and social justice issues. By leveraging media platforms, activists bring attention to important causes, humanizing environmental issues, and building public support for long-term solutions.

- *Media Attention*: Social movements frequently utilize various forms of media—whether traditional outlets like newspapers and television or newer platforms like social media—to amplify their messages. Through campaigns, press releases, and viral posts, activists can highlight environmental injustices, influence political discourse, and bring attention to problems that might otherwise be ignored. These media-driven efforts ensure that environmental issues remain part of public conversation.
- *Cultural Impact:* Beyond influencing legislation, social movements profoundly affect cultural norms and values. Over time, movements such as the climate justice or environmental justice movements help to shift societal perspectives, creating greater public awareness and encouraging broader acceptance of the need for sustainable practices. For example, the global push for action on climate change, fueled by movements like Fridays for Future, has led to greater recognition of the environmental crisis in both the media and public consciousness.

7.3.3.2. Empowering Communities

Grassroots activism not only focuses on broad systemic changes but also emphasizes empowering communities at the local level. By involving communities in decision-making processes and providing them with the necessary tools, resources, and knowledge, activists strengthen individuals' capacity to advocate for their own environmental and social rights.

- *Local Capacity Building*: One of grassroots activism's most important outcomes is its ability to foster local capacity building. This involves educating community members, developing their leadership skills, and providing access to resources so they can engage in environmental advocacy and decision-making processes. Empowered communities are better able to address their own local environmental challenges, from organizing clean-up initiatives to advocating for better waste management systems.
- *Strengthening Social Ties*: Grassroots movements also play a crucial role in fostering social cohesion. Community organizing efforts help to strengthen social ties, creating a sense of shared purpose and mutual support. This increased community collaboration leads to stronger, more resilient social networks, which are critical for tackling local environmental issues. Grassroots initiatives often emphasize collective action, enabling communities to approach challenges that affect them all (Stammers and Eschle, 2005; Tye et al., 2018; Leong et al., 2019).

7.3.4. Challenges Faced by Social Movements and Grassroots Activism

Grassroots organizations and social movements often face significant challenges related to resource limitations, which can affect their ability to operate effectively and achieve their objectives (Smith et al., 2014; Opiyo and Kuruç, 2016; Foweraker, 2001).

- *Funding:* A lack of financial resources is one of the primary constraints faced by grassroots movements. Many of these organizations rely on donations, grants, and fundraising efforts, which can be inconsistent or insufficient to meet the demands of sustained advocacy work. This limitation can restrict their ability to hire staff, implement projects, or run comprehensive campaigns that require materials, outreach, or travel.
- *Capacity:* Grassroots movements often operate with limited capacity in terms of technical support, infrastructure, and training. Many activists may lack access to the tools and expertise required to develop long-term strategies, evaluate project impacts, or engage in effective

lobbying. This can make it difficult for smaller, community-led initiatives to compete with well-funded organizations or powerful interest groups that have access to greater resources.

7.3.4.1. Resistance and Opposition

In the face of resistance and opposition, social movements frequently encounter barriers from powerful political and economic actors who seek to protect their own interests.

- *Political and Economic Interests*: Social movements often advocate for changes threatening established political or economic structures. This can lead to opposition from powerful actors, such as corporations or government entities, that have a vested interest in maintaining the status quo. These interests may use their influence to block reforms, discredit movements, or fund counter-campaigns designed to undermine activist efforts.
- *Repression:* In some cases, activists may face direct repression from authorities, including legal action, arrests, harassment, or even physical violence. Governments or corporations threatened by social movements may use these tactics to intimidate activists, disrupt organizing efforts, and weaken the impact of the movement. This kind of repression can stifle grassroots advocacy and discourage participation from community members.

7.3.4.2. Sustaining Momentum

Maintaining the momentum of social movements over time presents its own set of challenges, particularly as activists face personal and organizational pressures.

- *Burnout*: Activism often involves long hours, emotional labor, and a high level of dedication, leading to burnout among activists. This can be exacerbated when progress is slow or movements face repeated setbacks, leaving participants feeling overwhelmed, disillusioned, or fatigued. Without adequate support systems, activists may struggle to sustain their energy and commitment over the long term.
- *Maintaining Visibility*: As new issues arise and media attention shifts, it can be difficult for social movements to maintain visibility and keep their causes at the forefront of public discourse. Movements may lose momentum as interest wanes, particularly if they are not able to

generate ongoing engagement, evolve their strategies, or adapt to changing political landscapes. This makes it harder to sustain the level of public support and advocacy required for long-term change.

7.3.5. Successful Case Studies and Examples

The Green Belt Movement, founded by Wangari Maathai in Kenya, is a pioneering environmental initiative aimed at addressing deforestation, environmental degradation, and women's empowerment. Through a combination of tree planting and community organizing, the movement has led to the planting of over 50 million trees. This initiative not only helped combat deforestation and restore ecosystems but also significantly raised global awareness about environmental issues and highlighted the crucial role of women in conservation efforts.

The Dakota Access Pipeline Protests, centered on the Standing Rock Sioux Reservation in the United States, were driven by opposition to the construction of the Dakota Access Pipeline. The primary concern was the potential environmental impacts, particularly the threat to water sources, and the disregard for indigenous sovereignty. These protests brought significant national and international attention to indigenous rights and environmental justice, influencing public discourse and policy debates about resource extraction and environmental protection.

The Civil Rights Movement, led by figures like Martin Luther King Jr. and organizations such as the NAACP, was a critical social justice movement aimed at addressing racial segregation, discrimination, and broader social injustices in the United States. Through sustained activism and advocacy, the movement achieved significant legislative successes, including the Civil Rights Act of 1964 and the Voting Rights Act of 1965. These accomplishments not only advanced racial equality but also served as inspiration for subsequent social justice movements around the world.

The #MeToo Movement began as a social media campaign aimed at addressing sexual harassment and assault. Its primary goal was to raise awareness of the prevalence of these issues and to empower survivors to share their experiences and demand accountability. The movement quickly gained traction, sparking widespread conversations about gender equality, workplace harassment, and power dynamics. It has led to significant cultural and institutional changes, particularly in industries such as media, politics, and entertainment.

Social movements and grassroots activism are instrumental in driving environmental and social transformation. These efforts contribute to addressing injustices, promoting sustainability, and fostering social change by mobilizing communities, advocating for policy changes, and raising awareness (Lim, 2013; Mwangi and Njiir, 2021; Kostiner, 2003). Despite facing challenges, the impact of grassroots activism and social movements underscores the power of collective action and community engagement in shaping a more equitable and sustainable future.

7.4. Gender and Intersectionality in Environmental Policy

Gender and intersectionality are crucial dimensions in shaping environmental policy and understanding its impacts. These concepts highlight how various social identities and experiences intersect to influence individuals' interactions with the environment and their access to resources and opportunities. Incorporating gender and intersectionality into environmental policy is essential for creating equitable and effective solutions (Fletcher, 2018; Magnusdottir and Kronsell, 2021; Singleton et al., 2022).

7.4.1. Understanding Gender in Environmental Policy

7.4.1.1. Gender and Environmental Impact

Environmental issues tend to have differential impacts on men and women due to variations in gender roles, responsibilities, and access to resources. In many societies, women bear the primary responsibility for water collection, which makes them more vulnerable to issues such as water scarcity. This division of labor places a disproportionate burden on women when environmental resources become limited or degraded. Additionally, the health and safety risks posed by environmental hazards, such as exposure to pollution or climate-related disasters, can also differ between genders. Women and children, in particular, are often at greater risk due to their roles within households, including cooking and caregiving, which can increase their exposure to harmful substances and situations (Arora-Jonsson, 2014, 2017; Bhattacharyya and Rahman, 2020).

7.4.1.2. Gender-Inclusive Policies

Creating gender-inclusive policies in environmental management is critical to ensuring that diverse perspectives and experiences are incorporated into decision-making processes. Encouraging women's participation in these processes can lead to more comprehensive and effective policies, as they bring unique insights based on their roles in natural resource management. Furthermore, gender-sensitive policies that address the specific needs of both women and men are essential for ensuring equitable access to resources such as land, water, and technology. Capacity-building programs aimed at enhancing women's skills and knowledge in environmental management and leadership not only contribute to gender equality but also lead to better environmental outcomes by empowering women to take active roles in sustainable practices.

7.4.2. Intersectionality in Environmental Policy

7.4.2.1. The Concept of Intersectionality

Intersectionality, a concept coined by legal scholar Kimberlé Crenshaw, refers to how multiple social identities—such as race, gender, class, ethnicity, and disability—intersect to create overlapping systems of advantage or disadvantage. In the context of environmental policy, intersectionality is crucial for understanding how different forms of discrimination and inequality can influence individuals' experiences with environmental issues. For instance, a person's race, socioeconomic status, and gender might interact to create compounded environmental vulnerabilities. A low-income woman of color, for example, may face both the physical impacts of environmental degradation and the socio-economic challenges of marginalization, resulting in a complex web of disadvantage (Singleton et al., 2022; Fletcher, 2018; Magnusdottir and Kronsell, 2021).

7.4.2.2. Applying Intersectionality to Environmental Policy

7.4.2.2.1. Identifying Diverse Impacts

Environmental policies must be tailored to recognize the diverse ways in which identities and experiences intersect to shape environmental conditions and needs. By applying intersectionality, policymakers can create targeted solutions that address the specific challenges faced by different groups, ensuring that interventions are more inclusive and effective. Research and data collection are essential to this process; by reflecting a broad range of experiences and contexts, policymakers can develop a more comprehensive understanding of how environmental issues affect various communities.

7.4.2.2.2. Addressing Systemic Inequities

Intersectional environmental policies should focus on addressing systemic and historical inequities that have disproportionately impacted marginalized communities. Equitable distribution of resources, such as access to clean air, water, and green spaces, is essential for ensuring environmental justice. Additionally, policies should empower marginalized groups by enhancing their capacity to participate in environmental governance and decision-making. This involves creating opportunities for those most affected by environmental issues to have a voice in shaping solutions.

7.4.2.2.3. Integrating Diverse Perspectives

Engaging a wide range of stakeholders, including individuals from marginalized communities, is critical to developing inclusive environmental policies. By incorporating diverse perspectives, policymakers can better understand the needs and priorities of various groups and ensure that their voices are heard. Collaborative approaches that bring together different sectors, communities, and organizations can help create more inclusive and effective environmental policies, fostering a shared commitment to sustainability and justice.

7.4.3. Case Studies and Examples

7.4.3.1. Gender and Environmental Policy

7.4.3.1.1. The Women and Climate Change Agenda

Various international frameworks, such as the UNFCCC Gender Action Plan, underscore the importance of gender-responsive climate policies. These initiatives are designed to enhance women's participation in climate action and address the specific impacts of climate change on women. By incorporating gender considerations, policies become more comprehensive, leading to effective climate solutions that promote gender equality. The emphasis on women's roles in climate governance acknowledges their unique perspectives and contributions, ensuring that climate actions are inclusive and equitable (Buckingham and Le Masson, 2017; Castaneda et. al., 2012; Cohen, 2014).

7.4.3.1.2. The Women's Environmental Network (WEN)

The Women's Environmental Network (WEN) is an organization that integrates gender perspectives into environmental policy and activism. WEN advocates for women's rights and environmental justice, emphasizing the critical role of gender in promoting environmental protection and sustainability. By championing gender-sensitive approaches, WEN's work demonstrates how these strategies lead to more inclusive and equitable environmental policies, ensuring that the voices and needs of women are central to sustainability efforts.

7.4.3.2. Intersectionality in Environmental Policy

7.4.3.2.1. The Flint Water Crisis

The Flint water crisis serves as a stark example of how environmental injustice disproportionately affects marginalized communities. In Flint, the low-income Black population bore the brunt of the crisis, demonstrating the intersection of race, class, and environmental degradation. The crisis highlighted the need for policies that address racial and socio-economic disparities in environmental governance. As a result, efforts to improve infrastructure and ensure access to clean, safe water in marginalized communities have become key components of environmental policy responses.

7.4.3.2.2. Indigenous Land Rights and Environmental Protection

Indigenous communities frequently face intersectional challenges, including issues related to land rights, environmental degradation, and socio-economic marginalization. Policies recognizing Indigenous land rights and incorporating Indigenous knowledge systems are crucial for promoting environmental justice and effective protection. By involving Indigenous communities in environmental governance and respecting their cultural and ecological wisdom, policymakers can develop strategies that preserve natural resources and protect the rights of Indigenous peoples, contributing to more just and sustainable environmental outcomes.

7.4.4. Strategies for Integrating Gender and Intersectionality in Environmental Policy

A key step in creating effective environmental policies is conducting gender impact assessments. These assessments help evaluate how policies affect different genders, ensuring that they are responsive to gender-specific needs. By considering these nuances, policymakers can design interventions that address the unique challenges faced by women and men in relation to environmental issues. Additionally, using intersectional analysis is critical for understanding how various social identities, such as race, class, and gender, intersect to shape individuals' experiences. This approach ensures that policies are more comprehensive, addressing the multiple layers of inequality that different groups may face in environmental contexts (Singleton et al., 2022; Fletcher, 2018; Perkins, 2018).

7.4.4.1. Capacity Building and Education

To ensure that gender and intersectionality are effectively integrated into environmental policy, training programs should be developed for policymakers and practitioners. These programs would focus on educating stakeholders about the importance of considering gender and other intersecting identities in policy development and implementation. Furthermore, awareness campaigns are crucial in raising public and institutional understanding of how gender and intersectionality influence environmental outcomes. These campaigns can highlight the disproportionate effects of environmental issues on marginalized groups and promote more inclusive policy approaches.

7.4.4.2. Monitoring and Evaluation

For environmental policies to remain effective and equitable, inclusive indicators must be developed to measure their impact on different social groups. Data collection efforts should be designed to capture the diverse experiences of individuals from various backgrounds. This ensures that policies are not only inclusive in theory but also in practice. Additionally, conducting regular reviews of environmental policies allows for continuous assessment of their effectiveness in addressing gender and intersectional concerns. These evaluations are essential for identifying gaps and making necessary adjustments to improve equity and sustainability outcomes.

Incorporating gender and intersectionality into environmental policy is essential for creating equitable and effective solutions that address the diverse needs and experiences of all individuals. By recognizing the differential impacts of environmental issues and addressing systemic inequities, policymakers can develop more inclusive and just environmental policies that promote sustainability and social justice for all.

7.5. Public Participation and Community Empowerment

Public participation and community empowerment are fundamental components of effective environmental governance and policy development. These concepts ensure that communities have a meaningful role in decision-making processes and that their voices are heard in shaping policies that affect their environment and livelihoods (Kwok, 2000; Williams, 2018; Maina-Okori et al., 2018).

7.5.1. Understanding Public Participation

Public participation refers to the active involvement of individuals and communities in decision-making processes that directly affect their environment, health, and well-being. It involves various methods of engaging stakeholders, such as consultations, public hearings, surveys, and collaborative planning sessions. The importance of public participation lies in its ability to enhance transparency, accountability, and legitimacy in decision-making. When the public is involved, policies are more likely to reflect the real needs and preferences of the affected populations, leading to more effective, inclusive, and sustainable outcomes (Buchy and Hoverman, 2000; Quick and Bryson, 2022; Rowe and Frewer, 2004).

7.5.1.1. Levels of Public Participation

Public participation can occur at different levels, depending on the depth of engagement between the public and decision-makers.

- *Informing:* This is the most basic level, where the objective is to provide the public with balanced and objective information about policies, decisions, and their potential impacts. Common public information methods include reports, newsletters, websites, and public meetings. The purpose is to ensure that people know the decisions being made, even if they are not directly involved in shaping them.

- *Consulting*: At this level, the goal is to seek public feedback on policies, proposals, or plans before making final decisions. This can be achieved through surveys, focus groups, public forums, and consultations. While the public's input is considered, the final decision still lies with the authorities.
- *Involving*: Involving the public means actively including them in the decision-making process. Their input is not just collected but is seriously considered in shaping policies. Workshops, advisory committees, and collaborative planning sessions are common methods for involving the public.
- *Collaborating:* Collaboration takes participation further by engaging stakeholders in a process where they work alongside decision-makers to develop solutions. Methods include partnership agreements, co-design processes, and participatory planning. In this level, there is a shared responsibility in shaping policies and outcomes.
- *Empowering:* The highest level of public participation involves giving the public decision-making authority. This level ensures that the community has a direct role in shaping the outcomes of decisions. Methods include delegating decision-making powers to community groups, participatory budgeting, and community-led initiatives. Empowerment allows the public to influence the direction of policies that directly affect their lives.

7.5.2. Community Empowerment

Community empowerment refers to the process of enhancing the capacity and agency of communities to actively participate in decision-making, influence policies, and take action to address their own needs and concerns. This involves building the skills, confidence, and resources necessary for communities to engage in governance and problem-solving effectively. The importance of community empowerment cannot be overstated; empowered communities are better equipped to tackle local issues, advocate for their rights, and contribute to sustainable and equitable development. By fostering resilience, self-reliance, and collective action, community empowerment leads to stronger, more vibrant communities capable of driving positive change (Laverack, 2006; Shepard, 1993; Nikkhah and Redzuan, 2010).

7.5.2.1. Components of Community Empowerment

Community empowerment consists of several key components that work together to build the capacity of communities.

- *Capacity Building*: This foundational element includes providing education and training to community members, equipping them with the knowledge and skills needed to understand and engage in environmental governance. Additionally, ensuring access to financial resources, tools, and technologies is essential for supporting community initiatives. By enhancing capacity, communities become more effective in implementing their ideas and solutions.
- *Participation and Representation*: Inclusivity is crucial for effective community empowerment. It is important to ensure that all community members, especially marginalized groups, have opportunities to participate in and be represented in decision-making processes. Supporting the development of local leaders who can advocate for community interests is also vital, as these leaders can drive change and inspire collective action.
- *Advocacy and Mobilization*: Community empowerment encourages campaigns and advocacy efforts that community members themselves lead. Supporting these initiatives helps influence policies and raise awareness about environmental issues affecting the community. Additionally, facilitating collective action and collaboration among community members is essential for addressing shared challenges, fostering unity and purpose.
- *Self-Governance*: A critical aspect of community empowerment is encouraging and supporting local decision-making structures and processes. This allows communities to manage their own resources and address local issues more effectively. Promoting autonomy and self-determination in community development and environmental management reinforces the idea that communities know best what they need and how to achieve their goals. This empowerment leads to sustainable development aligned with the community's unique context and aspirations.

7.5.3. Strategies for Enhancing Public Participation and Community Empowerment

7.5.3.1. Creating Inclusive Platforms

To foster meaningful public engagement, it is essential to develop accessible and user-friendly channels that accommodate diverse community needs. This includes creating online platforms, organizing community meetings, and providing informational materials in multiple languages. Such accessibility ensures that all individuals, regardless of their backgrounds, can participate in discussions that impact their lives. Furthermore, implementing targeted outreach efforts is crucial for engaging underrepresented and marginalized communities. By actively reaching out to these groups, decision-makers can ensure that a wide range of voices and perspectives are heard, enriching the overall dialogue (Eklund, 1999; Blair, 2004; Adamson, 2010).

7.5.3.2. Building Trust and Relationships

Establishing trust between decision-makers and communities is vital for effective public participation. Maintaining transparency in decision-making fosters trust and accountability, as community members are informed about how their input will be used. Clear communication regarding decisions and their implications helps demystify the process and encourages ongoing engagement. Building strong relationships through regular communication and feedback also creates a supportive environment where communities feel valued and empowered to share their concerns and insights.

7.5.3.3. Supporting Capacity Building

To empower communities effectively, it is necessary to provide training programs that build essential skills in advocacy, policy analysis, and project management. These programs equip community members with the tools they need to engage confidently in environmental governance. Alongside training, resource allocation is crucial. Providing financial and technical support to community organizations and initiatives enhances their capacity to implement projects that address local needs and challenges, ensuring that communities are active participants in their development.

7.5.3.4. Promoting Collaborative Approaches

Collaboration is key to addressing environmental challenges effectively. Establishing partnerships between governments, non-governmental organizations (NGOs), businesses, and community organizations can lead to innovative solutions that benefit all stakeholders. By involving community members in the planning and implementation of environmental projects and policies, their needs and perspectives can be integrated into decision-making processes. This collaborative approach not only empowers communities but also fosters a sense of ownership and accountability for local initiatives.

7.5.3.5. Monitoring and Evaluation

To ensure the effectiveness of participation processes and policies, it is important to implement robust feedback mechanisms. These mechanisms enable communities to provide input on their experiences and the outcomes of participation efforts. By actively seeking feedback, decision-makers can identify areas for improvement and make necessary adjustments. Continuous improvement is vital; utilizing feedback and evaluation results allows for the refinement of participation and empowerment strategies, ensuring that they remain relevant and responsive to community needs over time.

7.5.4. Case Studies and Examples

7.5.4.1. Participatory Budgeting

Participatory budgeting is an innovative approach that empowers community members to have a direct say in how public funds are allocated. In cities like Porto Alegre, Brazil, and New York City, residents participate in processes that allow them to propose and vote on budgetary decisions. This model not only enhances public engagement but also promotes transparency in governmental operations. By involving citizens in financial decision-making, participatory budgeting fosters a sense of ownership and accountability among community members. The outcomes often reflect the diverse needs of the community, leading to more equitable resource distribution and improved public services.

7.5.4.2. Community-Led Conservation

Community-led conservation initiatives, such as Namibia's community-based natural resource management (CBNRM) program, exemplify the effectiveness of empowering local populations in environmental stewardship. This program enables communities to manage and benefit from their natural resources, particularly wildlife conservation efforts. By granting locals the authority and responsibility to protect their environment, the CBNRM has resulted in improved conservation outcomes while simultaneously enhancing community livelihoods. Moreover, it has strengthened local governance by fostering collaboration and decision-making at the community level, thereby ensuring that conservation efforts align with the needs and priorities of the residents.

7.5.4.3. Environmental Justice Movements

The Environmental Justice Movement in the United States highlights the critical intersection of environmental issues and social equity. Organizations such as the Sierra Club and the Center for Environmental Health advocate for the rights of marginalized communities to participate in environmental decision-making processes. This movement addresses the disproportionate environmental burdens faced by low-income and minority communities, ensuring that their voices are heard in discussions that impact their health and well-being. Through grassroots advocacy and policy initiatives, the Environmental Justice Movement has successfully led to significant policy changes and raised awareness about environmental justice issues, promoting a more inclusive approach to environmental governance.

Public participation and community empowerment are essential for effective and equitable environmental governance. By involving communities in decision-making processes and supporting their capacity to act, policies can be more responsive to local needs and challenges. Engaging diverse stakeholders, building trust, and fostering collaboration are key strategies for ensuring that environmental policies are inclusive, just, and effective (Barnes et al., 2007; Adamson, 2010; Fu and Ma, 2020).

Chapter 8

Challenges to Environmental Transformation

Environmental transformation involves profound changes to how societies interact with and impact their environment. Despite the growing recognition of the need for such transformation, several significant challenges hinder progress. These multifaceted challenges stem from socio-economic, political, technological, and cultural factors (Pugh, 2008; Feola, 2015; Elliot, 2011).

8.1. Political and Institutional Barriers

Political and institutional barriers are among the most significant challenges to environmental transformation. Many governments and institutions are influenced by short-term interests and economic considerations that can overshadow long-term environmental goals. Political instability, lack of commitment, and shifting priorities can impede the development and implementation of effective environmental policies. Additionally, bureaucratic inefficiencies and insufficient institutional capacity can hinder the enforcement of environmental regulations and the integration of sustainability into broader policy frameworks (Langlois-Bertrand et al., 2015; Misleh et al., 2024; Chang and Wu, 2014). In many cases, vested interests in industries such as fossil fuels and deforestation exert substantial pressure to maintain the status quo, further complicating efforts to enact meaningful change.

8.1.2. Economic Constraints

Economic constraints pose a significant challenge to environmental transformation. Transitioning to a sustainable economy often requires substantial financial investments in new technologies, infrastructure, and practices. For many developing countries and marginalized communities, the cost of adopting sustainable practices or transitioning to green technologies can be prohibitively high. Economic incentives and subsidies that support environmentally harmful practices, such as fossil fuel subsidies, can also create barriers to sustainability (Adabre et al., 2022; Shelomentsev et al., 2016; Rahman et al., 2020). The economic benefits of environmental transformation, such as long-term cost savings and health improvements, may not be immediately apparent, making it difficult to secure the necessary funding and support for transformational projects.

8.1.3. Technological Limitations

Technological limitations are another critical challenge. While technological innovation holds great promise for advancing environmental transformation, current technologies may not yet be sufficient to address all environmental challenges effectively. For example, while renewable energy technologies are advancing rapidly, issues related to energy storage, grid integration, and cost remain. Additionally, there are gaps in technologies needed for addressing complex problems like pollution control, waste management, and ecosystem restoration (Mazurkiewicz, 2018; Litvinenko et al., 2021; Mazurkiewicz, 20170). The development and deployment of cutting-edge technologies require significant research and development, which can be constrained by limited funding and resources.

8.1.4. Social and Cultural Resistance

Social and cultural resistance can also impede environmental transformation. Deeply ingrained social norms, practices, and lifestyles often create barriers to change. In many communities, traditional practices and economic activities are closely tied to environmental degradation, making it challenging to alter these behaviors without addressing underlying cultural values and economic dependencies. Resistance can also stem from misinformation or a lack of awareness about environmental issues and their impacts (Kiamba 2009; Oganisjana et al., 2017; Michael, 2019). Engaging communities in meaningful dialogue and education is essential to overcoming these barriers, but achieving broad social acceptance and behavioral change requires sustained effort and effective communication.

8.1.5. Lack of Global Coordination

Environmental issues are inherently global, yet efforts to address them often suffer from a lack of coordination and cooperation at the international level. Global environmental challenges, such as climate change and biodiversity loss, require coordinated action across borders, yet geopolitical conflicts, differing national interests, and varying levels of commitment can undermine international agreements and initiatives (Misleh et al., 2024; Kabra et al., 2015; Nazir and Sharifi, 2024). The complexity of global environmental issues necessitates robust and inclusive international frameworks, but achieving consensus among diverse stakeholders with differing priorities can be a major challenge.

8.1.6. Equity and Justice Concerns

Equity and justice concerns are crucial in the context of environmental transformation. Environmental policies and initiatives can disproportionately impact marginalized and vulnerable populations, exacerbating existing inequalities. For instance, the transition to green technologies might lead to job losses in traditional industries, affecting low-income workers disproportionately. Addressing these concerns requires careful consideration of social justice and equity in the design and implementation of environmental policies (Cecot and Hahn, 2024; Pereira et al., 2017; Schmitt and Maes, 2013). Ensuring that the benefits of environmental transformation are equitably distributed and that marginalized communities have a voice in decision-making processes is essential for achieving a just and inclusive transition.

8.1.7. Knowledge and Data Gaps

Knowledge and data gaps present significant obstacles to effective environmental transformation. Comprehensive and accurate data are essential for understanding environmental issues, tracking progress, and developing evidence-based policies. However, many regions lack reliable environmental data, and existing knowledge may be fragmented or outdated. This lack of information can hinder the development of effective strategies and limit the ability to monitor and evaluate the impacts of environmental policies (Slepchuk et al., 2022; Hughes-Hassell, 2020; Ward et al., 2018).

Strengthening data collection, research, and knowledge-sharing mechanisms is crucial for addressing these gaps and supporting informed decision-making.

8.1.8. Behavioral and Psychological Barriers

Behavioral and psychological barriers also play a role in hindering environmental transformation. Individuals and organizations may exhibit inertia or resistance to change due to cognitive biases, such as discounting future benefits or perceiving environmental issues as distant or abstract. Additionally, the complexity of environmental issues and the perceived difficulty of making a difference can lead to feelings of helplessness or apathy. Overcoming these barriers requires effective strategies to motivate and engage individuals, such as fostering a sense of agency, highlighting tangible benefits, and supporting sustainable behaviors (Alexander et al., 2024; Brown Travis and Compton, 2001; Tariq et al., 2020). Addressing the challenges to environmental transformation requires a multifaceted approach that encompasses political will, economic support, technological advancement, social engagement, global cooperation, equity considerations, knowledge development, and behavioral change. By recognizing and addressing these barriers, stakeholders can work towards creating a more sustainable and equitable future for all.

8.2. Technological Limitations and Ethical Concerns

Technological advancements are critical to addressing environmental challenges, but they also have limitations and ethical considerations. Understanding these aspects is essential for developing effective and responsible solutions (Marshall, 1999; Ronzhyn and Wimmer, 2019; Hua et al., 2024).

8.2.1. Technological Limitations

One of the significant technological gaps in the renewable energy sector is energy storage. As the adoption of renewable energy sources like solar and wind continues to rise, the need for efficient and cost-effective energy storage solutions becomes increasingly critical. Advanced battery technologies are essential for balancing supply and demand, as they enable the storage of excess energy generated during peak production times for use during periods of low generation. Without adequate energy storage solutions, the stability of the energy grid remains at risk, limiting the full potential of renewable energy integration (Ghaith, 2024; Baihakki and Qutayan, 2023; Šupa and Kruopštaitė, 2022).

Another pressing area of concern is waste management technology. While recycling processes have progressed, challenges persist in managing complex waste streams, particularly electronic and mixed plastics. The current recycling infrastructure is often ill-equipped to handle these materials effectively, leading to increased landfill use and environmental degradation. Additionally, there is a growing need for efficient waste-to-energy technologies that can reduce waste and lower greenhouse gas emissions. Developing innovative solutions for waste management is crucial to promoting a circular economy and mitigating environmental impacts.

Pollution control also presents a significant technological gap that requires urgent attention. Advanced air and water pollution control methods are necessary to tackle complex contaminants that pose risks to human health and the environment. Existing pollution control technologies may not adequately address emerging pollutants, such as microplastics and pharmaceuticals, which are identified as critical concerns. Ongoing research and innovation are essential to develop effective strategies for monitoring and mitigating pollution, ensuring a cleaner and healthier environment for future generations.

8.2.2. Economic and Practical Constraints

One of the primary economic constraints facing advanced environmental technologies is their high development and deployment costs. Technologies such as carbon capture and storage (CCS) hold great promise for reducing greenhouse gas emissions but come with substantial financial burdens. The significant investment required for research, development, and implementation can hinder widespread adoption, particularly in developing countries and among smaller organizations that may lack the necessary resources. As a result, the financial barriers associated with advanced environmental technologies can slow progress in mitigating climate change and improving sustainability (Higgott, 1993; Kuo, 2008; Sulaiman et L., 2012).

In addition to high costs, scalability presents a practical challenge for many innovative technologies. While some new technologies may prove effective on a small scale, expanding their application for widespread use often proves difficult. A prime example is electric vehicles (EVs), which are becoming increasingly popular as a sustainable transportation option. However, the infrastructure needed to support EVs, including charging stations, has not yet been fully developed to meet the growing demand. Furthermore, the supply chains for critical materials used in batteries face constraints that complicate efforts to scale production. Addressing these scalability issues is essential to facilitate the broader adoption of advanced environmental technologies and maximize their potential impact on sustainability and climate resilience.

8.2.3. Technological Risks and Unintended Consequences

While many technologies are developed to address environmental challenges, they can sometimes have unintended negative consequences. A notable example is large-scale biofuel production, which, if not managed sustainably, can lead to significant issues such as deforestation and biodiversity loss. The conversion of forests and natural habitats into biofuel plantations disrupts ecosystems, threatens wildlife, and contributes to the environmental degradation these technologies aim to combat. Similarly, geoengineering techniques, intended to mitigate climate change by altering the Earth's systems, can introduce unforeseen impacts on ecosystems and weather patterns (Pringle et al., 2016; Kuziemsky et al., 2016; Miller and Engemann, 2019). These interventions may inadvertently destabilize local climates, disrupt food systems, and affect water availability, highlighting the need for careful consideration and robust management when implementing such technologies.

8.2.3.1. Dependency on Technology

Another critical concern is the potential over-reliance on technological solutions, which can distract from the necessary behavioral and systemic changes for achieving true sustainability. While innovations can provide valuable tools for addressing environmental issues, focusing exclusively on technological fixes may overshadow the importance of altering consumption patterns and energy use behaviors. For instance, the belief that advanced technologies alone can solve climate change might lead to complacency regarding necessary lifestyle changes, such as reducing waste, conserving energy, and shifting towards more sustainable practices (Vermaas et al., 2011; Kendal, 2022; Hillerbrand, 2010). Addressing these underlying issues requires a holistic approach integrating technological advancements with social, economic, and behavioral changes, ensuring a more comprehensive and effective response to environmental challenges.

8.2.4. Ethical Concerns

8.2.4.1. Equity and Access

Technological advancements can often exacerbate existing inequalities, particularly when access to new technologies is limited to wealthier individuals or countries. For example, advanced water purification technologies might be widely available in affluent regions, providing clean and safe drinking water to those populations, while poorer areas may lack access to such essential resources. This disparity not only impacts health and quality of life but also widens the gap between different socio-economic groups, perpetuating cycles of poverty and disadvantage. If technological progress is not equitably distributed, it risks reinforcing systemic inequities, leaving vulnerable populations further behind in their ability to respond to environmental challenges (Chen et al., 2019; Ategeka, 2021; Sulecio de Alvarez and Dickson-Deane, 2018).

The implementation of technological solutions can have varied impacts on different communities, particularly vulnerable ones. For instance, large-scale renewable energy projects, such as wind farms or solar arrays, may result in the displacement of local communities or significant changes to their land use. In many cases, these projects are developed with insufficient consultation or compensation for those affected, leading to loss of livelihoods and cultural displacement. Additionally, without adequate consideration of local needs and contexts, such initiatives can create tensions and conflicts between communities and developers. It is essential to ensure that technological solutions not only address environmental issues but also include the voices and concerns of affected communities, fostering equitable development and respect for local rights.

8.2.4.2. Privacy and Surveillance

Technologies used for environmental monitoring and management often involve collecting large amounts of data. This raises concerns about data privacy and the potential misuse of information. For example, smart grids and environmental sensors might track personal data that could be exploited if not properly safeguarded (Baggett, 2017; Cavoukian et al., 2014).

Using drones for environmental monitoring can raise surveillance and privacy issues. Balancing the benefits of data collection for environmental protection with individuals' rights to privacy is a significant ethical challenge.

8.2.4.3. Informed Consent and Autonomy

Implementing new technologies, especially those involving genetic modifications or geoengineering, raises questions about informed consent. Affected populations must be adequately informed and have the opportunity to consent to or reject technologies that impact their environment or health (Tan et al., 2021; Hung et al., 2022; O'Meara and O'Meara, 2014).

Technological interventions can sometimes override local knowledge and control. For example, introducing new technologies in Indigenous lands without their consent can undermine traditional practices and local autonomy. Ensuring communities have a say in technological decisions is essential for respecting their rights and preferences.

8.2.4.4. Long-Term Risks and Uncertainty

Many technological interventions carry long-term risks and uncertainties. For instance, the long-term impacts of climate engineering methods, such as solar radiation management, are not fully understood and could have unintended consequences for global ecosystems and climate systems.

Deciding which technologies to develop and deploy involves ethical considerations about potential benefits and risks. To make informed and responsible decisions, stakeholders must carefully weigh the potential positive impacts against the risks and ethical implications.

Technological limitations and ethical concerns are integral to the discourse on environmental transformation. Addressing technological gaps requires ongoing research, innovation, and investment, while ethical considerations demand a careful balancing of equity, privacy, and community rights (Tonn and Stiefel, 2019; Miller and Engemann, 2019; Healy, 2012). By acknowledging and addressing these challenges, stakeholders can work towards solutions that are not only technologically advanced but also socially responsible and equitable.

8.3. Environmental Degradation and Resource Scarcity

Environmental degradation and resource scarcity are pressing issues that significantly challenge global sustainability and human well-being. They are interconnected phenomena that result from the unsustainable exploitation of natural resources and the degradation of ecosystems (Dinar, 2011, 2011a; Lonergan, 1998). Understanding these issues is crucial for developing effective strategies to address them and promote long-term environmental health.

8.3.1. Environmental Degradation

Environmental degradation refers to the deterioration of the natural environment due to human activities. This degradation manifests in various forms, including deforestation, soil erosion, water pollution, and biodiversity loss (Mohamed and Nageye, 2019; Elisha, 2021; van den Bergh, 2012).

8.3.1.1. Deforestation

Deforestation, or the large-scale removal of forests, has profound impacts on ecosystems and the climate. Forests are crucial for regulating the global climate, as they absorb carbon dioxide and release oxygen. The loss of forests disrupts these processes, contributing to climate change. Additionally, deforestation leads to habitat loss, endangering countless species of plants and animals (Jha, 2018; Sirin, 2011; Loskutova, 2020). The destruction of forest ecosystems also affects indigenous communities who rely on forests for their livelihoods and cultural practices.

8.3.1.2. Soil Erosion and Degradation

Soil erosion and degradation result from unsustainable agricultural practices, deforestation, and urbanization. Erosion removes the nutrient-rich topsoil necessary for crop growth, reducing agricultural productivity and leading to food insecurity. Soil degradation also affects water quality, as eroded soil can carry pollutants into waterways, causing sedimentation and affecting aquatic ecosystems (Issaka and Ashraf, 2017; De Ploey, 1991; Lal, 2001). Sustainable land management practices are essential to prevent soil degradation and maintain soil health.

8.3.1.3. Water Pollution

Water pollution occurs when harmful substances contaminate water bodies, affecting their quality and the health of aquatic life. Pollutants such as chemicals, heavy metals, and plastics enter rivers, lakes, and oceans through industrial discharges, agricultural runoff, and improper waste disposal. Water pollution poses significant risks to human health, as contaminated water can cause diseases and disrupt ecosystems (Chaudhry and Malik, 2017; Dwivedi, 2017; Moss, 2008; Goel, 2006). Protecting water resources requires stringent regulations on pollutants, improved waste management, and the adoption of cleaner technologies.

8.3.1.4. Loss of Biodiversity

The loss of biodiversity, or the reduction in the variety of life forms within an ecosystem, is a critical aspect of environmental degradation. Human activities such as habitat destruction, pollution, and climate change contribute to species extinction and the decline of populations. Biodiversity loss undermines ecosystem resilience, reducing the ability of ecosystems to provide essential services such as pollination, water purification, and climate regulation (Smith et al., 2003; Cardinale et al., 2012; Robinson, 1993; Bhandari 2023). Conservation efforts are crucial to preserving biodiversity and maintaining ecosystem health.

8.3.2. Resource Scarcity

Resource scarcity refers to the depletion of natural resources due to overuse and mismanagement. It encompasses a range of resources, including water, minerals, fossil fuels, and arable land (Krautkraemer, 1998; Cannon et al., 2019).

8.3.2.1. Water Scarcity

Water scarcity is a growing concern, affecting both the availability and quality of freshwater resources. Population growth, climate change, and pollution exacerbate water scarcity issues. In many regions, over-extraction of water for agriculture, industry, and domestic use depletes aquifers and reduces river flows. Climate change also impacts water availability by altering precipitation patterns and increasing the frequency of droughts (Liu et al., 2017; Seckler et al., 1999). Addressing water scarcity requires integrated water management strategies, conservation efforts, and investments in water infrastructure.

8.3.2.2. Mineral and Fossil Fuel Depletion

The extraction and consumption of minerals and fossil fuels contribute to resource depletion and environmental degradation. Fossil fuels such as coal, oil, and natural gas are finite resources that, when burned, release greenhouse gases and pollutants, exacerbating climate change and air quality issues. Mineral extraction, including mining for metals and rare earth elements, can lead to habitat destruction, soil erosion, and water contamination. Transitioning to renewable energy sources and adopting circular economy principles can help mitigate the impacts of resource depletion (Whiting et al., 2017; Shafiee and Topal, 2009).

8.3.2.3. Arable Land and Food Security

The scarcity of arable land suitable for agriculture poses a challenge to food security. Urbanization, deforestation, and soil degradation reduce the amount of land available for farming. Additionally, intensive agricultural practices can deplete soil nutrients and reduce land productivity. Ensuring food security requires sustainable agricultural practices, land conservation, and innovations in food production and distribution (Chen et al., 2019; Nath et al., 2015; Gardi et al., 2015).

8.3.2.4. The Circular Economy Approach

The circular economy offers a framework for addressing resource scarcity by promoting the reuse, recycling, and sustainable management of resources. Unlike the traditional linear economy, which follows a "take-make-dispose" model, the circular economy aims to close the loop by keeping resources in use for as long as possible and minimizing waste. By redesigning products, improving recycling processes, and encouraging sustainable consumption, the circular economy can help alleviate resource scarcity and reduce environmental impacts (Hong and Chen, 2017; Stahel, 2016).

Environmental degradation and resource scarcity are interrelated issues that require comprehensive and coordinated responses. Addressing these challenges involves promoting sustainable practices, protecting ecosystems, and managing resources efficiently. By adopting innovative solutions, implementing effective policies, and fostering global cooperation, we can work towards a more sustainable and resilient future for both people and the planet.

8.4. Global Inequality and Access to Resources

Global inequality and access to resources are critical issues that profoundly impact the effectiveness of environmental strategies and the sustainability of development efforts. These disparities not only exacerbate existing social and economic divides but also hinder progress toward equitable and sustainable global solutions (Robeyns, 2017; Altmann et al., 2013; Babones and Turner, 2004).

8.4.1. Disparities in Resource Availability

Economic inequality significantly affects access to resources. Wealthier countries and individuals typically have greater access to essential resources such as clean water, energy, and food. In contrast, poorer regions often face shortages and limited access. This disparity is reflected in the uneven distribution of natural resources, where some countries with abundant resources still experience poverty and deprivation due to poor management or exploitation by wealthier nations (Jones et al., 2015; Flaskerud, 2022; Horev et al., 2004).

Access to advanced infrastructure and technology further exacerbates resource inequality. Developed countries often have the infrastructure needed to manage and utilize resources efficiently, while developing countries may lack such infrastructure, leading to inefficient use and higher levels of waste and environmental degradation. For example, advanced water purification and recycling technologies are more accessible in high-income countries, whereas many low-income regions struggle with basic sanitation and clean water access.

8.4.2. Impacts on Human Health and Well-Being

Inequitable access to resources directly impacts health outcomes. In regions with limited access to clean water, adequate sanitation, and nutritious food, communities are more vulnerable to health issues such as waterborne diseases, malnutrition, and other preventable illnesses. The lack of resources affects physical health and mental well-being, contributing to broader socio-economic challenges (Brusseau et al., 2019; Corvalan et al., 2005; Butler et al., 2005).

Access to education, healthcare, and energy resources is crucial in determining quality of life. Disparities in these areas lead to differences in educational attainment, economic opportunities, and overall life satisfaction. For instance, energy poverty limits access to modern amenities and services, affecting productivity and quality of life in affected regions.

8.4.3. Environmental and Social Implications

Inequitable access to resources can lead to environmental degradation. In regions where resources are scarce or mismanaged, overexploitation and unsustainable practices often occur. For example, in some developing countries, limited access to modern agricultural technologies leads to practices that can degrade soil quality and reduce agricultural productivity over time (Williams et al., 2008; Obidzinski et al., 2012).

Resource inequality can contribute to social and political tensions. Disparities in resource access can lead to conflicts over resources, migration pressures, and social unrest. The competition for limited resources can exacerbate existing inequalities and lead to conflicts both within and between countries.

8.4.4. Global Environmental Justice

Global environmental justice considers the historical responsibility of wealthier nations for environmental degradation and resource depletion. Many high-income countries have historically contributed the most to environmental problems such as greenhouse gas emissions and deforestation, while low-income countries often bear the brunt of these issues despite having contributed minimally (Martinez-Alier et al., 2016; Temper et al., 2015).

Ensuring fair distribution of resources requires addressing the root causes of inequality and implementing policies that promote equitable access. This includes international cooperation, support for development initiatives, and strategies that consider the needs of the most vulnerable populations.

8.4.5. Policy and Governance

International agreements and frameworks often facilitate global efforts to address resource inequality and environmental challenges. Agreements such as the Sustainable Development Goals (SDGs) aim to promote equitable access to resources and address global inequalities. These frameworks emphasize the need for collaborative approaches to tackle resource scarcity and ensure that development benefits are distributed fairly (Farazmand, 2023; Craft and Howlett, 2012).

At the national and local levels, policies that promote resource equity and sustainable development are essential. This includes investing in infrastructure, supporting local economies, and implementing social safety nets. Policies should be designed to address the specific needs of marginalized and vulnerable communities, ensuring that they have the resources and support needed to thrive.

8.4.6. Sustainable Development and Innovation

Promoting inclusive innovation can help bridge the resource gap. Technologies and practices adaptable to different contexts and accessible to underserved communities can be crucial in addressing resource disparities. For instance, low-cost water purification technologies and renewable energy solutions can significantly improve resource access in low-income regions.

Empowering local communities and integrating their knowledge and needs into resource management can lead to more equitable and effective solutions. Community-led initiatives and participatory approaches ensure solutions are tailored to local contexts and contribute to long-term sustainability (Salam, 2024; Vollenbroek, 2002; Szopik-Depczyńska et al., 2018).

Global inequality and access to resources are deeply intertwined issues that impact environmental sustainability and social equity. Addressing these challenges requires a multifaceted approach that includes international cooperation, equitable policy-making, and innovative solutions. By focusing on fair resource distribution and empowering vulnerable communities, it is possible to promote a more just and sustainable world for all.

Chapter 9

Environmentalism in a Globalized World

Environmentalism in a globalized world represents the intersection of ecological concerns with the economic, social, and political dimensions of global interconnectedness. Globalization has accelerated the flow of goods, services, people, and ideas across borders, contributing to economic growth and development. However, it has also heightened the environmental challenges the world faces, such as resource depletion, climate change, and biodiversity loss (Luke, 2005; Schofer and Granados, 2006; Curran, 2006). As a result, environmentalism in the context of globalization must contend with the complexities of a world where local actions have global repercussions and vice versa.

9.1. The Global Nature of Environmental Problems

One of the central tenets of environmentalism in a globalized world is the recognition that many environmental challenges are transnational in scope. Climate change, for example, is a global issue that affects every region on the planet, albeit in different ways. The burning of fossil fuels in one part of the world can contribute to rising sea levels, extreme weather events, and shifting ecosystems elsewhere. Similarly, the destruction of tropical rainforests has far-reaching consequences for global biodiversity and the earth's capacity to absorb carbon dioxide (Singh and Singh, 2017; Taylor and Buttel, 1992; Owen et al., 2006).

The interconnectedness of the global economy means that environmental degradation in one country can be driven by consumption patterns in another. Deforestation in the Amazon, for instance, is often linked to global demand for beef, soy, and palm oil. This creates a complex web of environmental impacts where local ecosystems are affected by global supply chains, requiring environmentalism to adopt a broader, more integrated approach that addresses both local actions and their global drivers.

9.1.1. The Role of International Environmental Governance

In response to the global nature of environmental challenges, international organizations and treaties have become critical platforms for coordinating environmental action. Bodies like the United Nations Environment Programme (UNEP) and international agreements such as the Paris Agreement under the United Nations Framework Convention on Climate Change (UNFCCC) provide frameworks for countries to collaborate on addressing issues such as climate change, biodiversity loss, and pollution.

The global nature of these agreements reflects the understanding that no single country can tackle these problems alone. However, international environmental governance also faces significant challenges, such as differing national interests, economic disparities between nations, and varying levels of political will. While wealthy nations often have the financial resources and technology to implement environmental policies, developing countries may struggle to balance environmental protection with economic growth (Young, 2003; Bernstein, 2004; Zürn, 1998). Global environmental governance must therefore address issues of fairness, equity, and capacity building to ensure that all nations can contribute to and benefit from environmental solutions.

9.1.2. The Environmental Impacts of Global Trade and Consumption

Global trade has expanded dramatically in recent decades, driven by the liberalization of markets and advancements in transportation and communication technologies. While trade has brought economic benefits and lifted millions out of poverty, it has also contributed to significant environmental degradation. The production and transportation of goods across vast distances consumes large amounts of energy, particularly from fossil fuels, contributing to greenhouse gas emissions. Moreover, the global demand for natural resources, such as timber, minerals, and oil, often leads to the overexploitation of ecosystems, particularly in developing countries that rely on exports for economic growth (Wiedmann et al., 2007; Chaudhary and Brooks, 2019; Balogh and Jámbor, 2020).

The rise of consumerism in a globalized world also plays a significant role in environmental degradation. The demand for cheap, disposable goods often leads to unsustainable production practices that deplete resources and generate waste. Environmentalists argue for more sustainable consumption patterns, emphasizing the need for responsible production, the reduction of waste, and the promotion of circular economy models. Addressing the environmental impacts of global trade and consumption requires concerted efforts to shift towards more sustainable production practices and consumer behaviors on a global scale.

9.1.3. Environmental Justice in a Globalized World

Globalization has also highlighted disparities in how environmental harms and benefits are distributed across the world. Often, the environmental costs of globalization—such as pollution, deforestation, and the loss of biodiversity—are borne disproportionately by poorer countries and communities. At the same time, wealthier nations and corporations reap the economic rewards. This has led to a growing call for environmental justice on a global scale, where the rights of marginalized communities, indigenous peoples, and the Global South are acknowledged and protected in environmental decision-making processes (Roberts, 2007; Reed and George, 2011; Mohai et al., 2009).

For example, small island nations that contribute little to global greenhouse gas emissions are among the most vulnerable to the impacts of climate change, such as rising sea levels and extreme weather events. Similarly, indigenous communities that have historically stewarded natural ecosystems are often displaced or marginalized by industrial development. Environmental justice movements advocate for the equitable distribution of environmental benefits and burdens, pushing for policies that protect vulnerable populations from environmental harms and ensure their participation in decision-making.

9.1.4. The Role of Technology and Innovation in Global Environmentalism

Technology plays a dual role in environmentalism in a globalized world. On the one hand, technological advancements—such as the widespread use of fossil fuels, intensive agriculture, and industrial manufacturing—have driven many of the environmental problems we face today. On the other hand, technology and innovation are also key to solving these problems. The development of renewable energy technologies, such as solar, wind, and hydropower, offers the potential to decarbonize energy systems and reduce reliance on fossil fuels. Likewise, advances in environmental monitoring, such as satellite imagery and big data analytics, provide new tools for tracking deforestation, pollution, and climate change impacts in real time (Jansen and Vellema, 2008; Jamison, 1996; Shellenberger and Nordhaus, 2009).

Moreover, globalization has facilitated the rapid spread of environmental knowledge and solutions across borders. Environmental movements, once localized, now have a global reach thanks to the internet and social media. International collaboration on scientific research has accelerated the development of new technologies and practices that can mitigate environmental harms, such as carbon capture and storage, sustainable agriculture, and water conservation techniques.

9.1.5. Globalized Environmental Activism

Environmentalism in a globalized world has also given rise to transnational environmental movements. Activists and NGOs now operate across borders, advocating for global environmental action and mobilizing citizens around the world. Social media has played a crucial role in amplifying these movements, allowing environmental campaigns to gain global attention rapidly. Movements like Fridays for Future, led by young climate activists such as Greta Thunberg, have transcended national boundaries, highlighting the global nature of the climate crisis and calling for urgent, coordinated action from governments around the world (Wapner, 1996; Ignatow, 2008).

This globalized environmental activism also emphasizes the interconnectedness of social, economic, and environmental issues. Many environmental movements today incorporate concerns about social justice, human rights, and economic inequality, recognizing that environmental sustainability cannot be achieved without addressing these broader systemic issues.

Environmentalism in a globalized world must navigate the complexities of an interconnected planet, where local actions reverberate globally and vice versa. It requires a coordinated, multi-level approach that involves international cooperation, technological innovation, and a commitment to environmental justice. As globalization continues to shape our world, the challenge for environmentalism is to harness the benefits of interconnectedness while mitigating the environmental harms that often accompany global development. Only by recognizing the global nature of these challenges can we hope to achieve a sustainable and equitable future for all.

9.2. Globalization, Trade, and Environmental Impacts

Globalization, characterized by the increasing interconnectedness and interdependence of economies, cultures, and societies, has profound implications for the environment. The expansion of international trade, a key aspect of globalization, has both positive and negative impacts on environmental sustainability. Understanding these impacts requires examining how global trade practices affect natural resources, ecosystems, and climate change, and exploring potential strategies for mitigating adverse effects while maximizing benefits (Lofdahl, 2002; Rehman et al., 2021; Ma and Wang, 2021).

9.2.1. Positive Environmental Impacts of Globalization and Trade

Globalization facilitates the spread of technological innovations and best practices across borders. Advanced environmental technologies, such as renewable energy systems, pollution control technologies, and waste management solutions, can be disseminated globally through trade. For example, the adoption of solar and wind energy technologies has been accelerated by international trade, contributing to cleaner energy production and reduced greenhouse gas emissions (Lofdahl, 2002; Panayotou, 2000; Balogh and Jámbor, 2020).

Global trade allows for economies of scale, which can lead to more efficient production processes and resource use. Large-scale production can result in lower per-unit environmental impacts compared to smaller-scale operations. For instance, the consolidation of manufacturing in regions with stringent environmental regulations can reduce the overall environmental footprint of products by minimizing waste and optimizing resource use.

9.2.2. Negative Environmental Impacts of Globalization and Trade

The demand for raw materials driven by global trade can lead to overexploitation of natural resources. Countries with abundant resources, such as forests, minerals, and fossil fuels, may experience significant environmental degradation as they extract and export these resources to meet global demand. Deforestation, soil erosion, and habitat destruction are common consequences of resource extraction activities (Baek et al., 2009; Balogh and Jámbor, 2020; Figge et al., 2017).

Global trade often involves the movement of goods across long distances, increasing transportation-related pollution. The shipping and aviation sectors contribute to greenhouse gas emissions, air pollution, and marine pollution. Additionally, the export of waste and hazardous materials from high-income countries to developing countries can result in environmental and health risks, as these regions may lack adequate waste management infrastructure.

The global supply chain can increase the carbon footprint of products due to the energy and emissions associated with transportation, manufacturing, and processing. Products manufactured in one country and consumed in another may have a higher environmental impact compared to locally produced goods. This "carbon leakage" can undermine efforts to reduce overall greenhouse gas emissions.

9.2.3. Trade Policies and Environmental Standards

Trade policies and agreements can influence environmental outcomes by setting standards and regulations that govern international trade. For instance, trade agreements can include environmental provisions that require countries to adhere to certain environmental standards, promote sustainable practices, and prevent the degradation of natural resources. The inclusion of environmental clauses in trade agreements, such as those found in the Comprehensive and Progressive Agreement for Trans-Pacific Partnership (CPTPP), reflects the growing recognition of the need to integrate environmental considerations into trade policy (Anderson, 1996; Barrett, 2001; Bastiaens and Postnikov, 2017).

Some countries have implemented border carbon adjustments or tariffs to address the environmental impacts of imported goods. These measures aim to level the playing field by imposing additional costs on imports with higher carbon footprints, encouraging foreign producers to adopt more sustainable practices. Such policies can help mitigate the environmental impacts of global trade by incentivizing lower-carbon production and reducing carbon leakage.

9.2.4. Sustainable Trade and Consumption

The circular economy concept promotes sustainable trade and consumption by emphasizing the reuse, recycling, and recovery of materials. By extending the life cycle of products and reducing waste, the circular economy can help minimize the environmental impacts associated with global trade. Encouraging circular practices, such as designing products for durability and recyclability, can reduce the demand for raw materials and lower the carbon footprint of traded goods (Zaccaï, 2007; Jackson, 2014).

Sustainable sourcing practices involve selecting materials and products that are produced in an environmentally and socially responsible manner. This includes prioritizing certified sustainable products, such as those with Forest Stewardship Council (FSC) certification for timber or Fair Trade certification for agricultural products. By promoting sustainable sourcing, businesses and consumers can contribute to reducing the environmental impact of global trade.

9.2.5. Global Cooperation and Environmental Governance

Addressing the environmental impacts of globalization and trade requires international cooperation and governance. Multilateral agreements, such as the Paris Agreement on climate change and the Convention on Biological Diversity, are crucial in setting global environmental standards and fostering cooperation among countries. These agreements aim to harmonize efforts to address environmental challenges and promote sustainable development.

Global initiatives, such as the United Nations Sustainable Development Goals (SDGs), provide a framework for integrating environmental considerations into trade and economic policies. The SDGs emphasize the importance of sustainable consumption and production, climate action, and the protection of ecosystems. By aligning trade policies with these global goals, countries can work together to promote environmental sustainability in a globalized world.

Globalization and trade have significant environmental implications, with both positive and negative impacts on natural resources, ecosystems, and climate change. While globalization can facilitate the spread of sustainable technologies and promote efficient resource use, it also presents challenges such as resource overexploitation, pollution, and increased carbon footprints. Addressing these challenges requires a multifaceted approach that includes implementing effective environmental regulations, promoting sustainable trade practices, and fostering international cooperation (Biermann and Pattberg, 2012; Bernstein, 2004; Esty, 2008). Integrating environmental considerations into global trade policies and practices makes it possible to balance economic growth and environmental sustainability, ensuring a healthier planet for future generations.

9.3. Cultural Perspectives on Environmentalism

Cultural perspectives on environmentalism provide valuable insights into how different societies understand and interact with the natural world. Environmentalism, as a global movement, is shaped by diverse cultural values, beliefs, and practices that influence environmental attitudes and actions. Exploring these cultural perspectives helps to understand the varied approaches to environmental stewardship and highlights the importance of incorporating cultural diversity into global environmental strategies (Grendstad et al., 2006; Boyle, 2007).

9.3.1. Indigenous Perspectives and Traditional Ecological Knowledge

Indigenous cultures often possess a holistic worldview that emphasizes the interconnectedness of humans, nature, and the spiritual realm. Traditional ecological knowledge (TEK) developed over generations reflects a deep understanding of local ecosystems, natural processes, and sustainable resource management practices. For example, many Indigenous communities practice rotational farming, sustainable hunting, and sacred site protection, which are rooted in a respect for nature and the recognition of its limits (Dudgeon and Berkes, 2003; Isaac et al., 2018).

Indigenous cultural practices and rituals often incorporate conservation principles. Sacred landscapes, totemic species, and spiritual ceremonies play a role in maintaining ecological balance and fostering respect for the environment. Recognizing and valuing TEK can enhance contemporary conservation efforts by integrating traditional wisdom with scientific knowledge, leading to more effective and culturally sensitive environmental management.

9.3.2. Eastern Philosophies and Environmentalism

Confucianism and Harmony- Confucianism, a major philosophical tradition in East Asia, emphasizes harmony between humans and nature. This perspective advocates for a balanced relationship with the environment, where human actions are guided by principles of moderation, respect, and benevolence. The concept of "Tianxia" (all under heaven) reflects the idea of a unified and harmonious world, encouraging practices that protect natural resources and promote ecological balance (Callicott and McRae, 2014; Droz et al., 2024).

Buddhism and Interconnectedness- Buddhism, with its focus on the interdependence of all living beings, promotes environmental mindfulness. The principle of "ahimsa" (non-harm) extends to the natural world, advocating for a compassionate approach to environmental issues. Buddhist practices, such as mindfulness and simplicity, contribute to environmental conservation by encouraging individuals to live in harmony with nature and reduce their ecological footprint (Bhandari, 2019, 2019a; 2022).

9.3.3. Western Environmentalism and Modernity

Industrialization and Environmental Awareness- In the Western context, environmentalism has often emerged in response to the environmental impacts of industrialization and urbanization. The rise of the modern environmental movement in the 20th century was marked by a growing awareness of pollution, resource depletion, and ecological degradation. Key milestones include the publication of Rachel Carson's "Silent Spring," which highlighted the dangers of pesticides, and the establishment of environmental organizations such as the Sierra Club and Greenpeace (Argyrou, 2005).

Consumerism and Sustainability- Western consumer culture has influenced environmentalism by driving the demand for sustainable products and practices. The rise of the "green consumer" and movements such as "zero waste" reflect a growing awareness of the environmental impacts of consumption. However, the challenge remains to reconcile material abundance's values with sustainability and conservation principles (Meyer, 2001).

9.3.4. African Perspectives and Environmental Stewardship

The Ubuntu philosophy, prevalent in many African cultures, emphasizes community, interconnectedness, and mutual respect. This perspective fosters a sense of responsibility toward the environment, recognizing that the well-being of individuals is intertwined with the health of the natural world. Traditional African practices, such as community-based resource management and respect for sacred natural sites, reflect this holistic approach to environmental stewardship (Mugambi, 2016; Barendse et al., 2016).

Resource Management and Traditional Practices- African communities often employ traditional practices for sustainable resource management, such as agroforestry, water conservation techniques, and rotational grazing. These practices are guided by a deep understanding of local ecosystems and a cultural commitment to preserving natural resources for future generations (Darko, 2014).

9.3.5. Latin American Perspectives and Environmental Justice

Pachamama and Andean Cosmology- In many Latin American cultures, particularly among Indigenous Andean peoples, Pachamama (Mother Earth) represents a sacred and life-giving entity. This cosmology emphasizes the reciprocal relationship between humans and the earth, advocating for respect, gratitude, and sustainable use of natural resources. Rituals and ceremonies honoring Pachamama reflect a cultural commitment to environmental conservation (Álvarez and Coolsaet, 2020).

Environmental Justice Movements- Latin American countries have seen the rise of environmental justice movements that address the social and environmental impacts of development projects, such as mining and deforestation. These movements highlight the intersection of environmental and social justice, advocating for the rights of marginalized communities and protecting natural resources from exploitation (Ulloa, 2017; Rodriguez, 2020).

9.3.6. Cultural Diversity and Global Environmentalism

9.3.6.1. Integrating Cultural Values

Recognizing and integrating cultural values into global environmental strategies is essential for achieving effective and equitable solutions. Cultural diversity enriches environmentalism by offering diverse conservation, resource management, and sustainability approaches. Collaborating with communities to understand their cultural perspectives and practices can enhance environmental policies and foster more inclusive and respectful approaches to environmental stewardship (Falkner, 2021).

9.3.6.2. Challenges and Opportunities

Balancing diverse cultural perspectives with global environmental goals presents both challenges and opportunities. Ensuring global environmental initiatives respect and incorporate cultural diversity requires open dialogue, collaboration, and sensitivity to local contexts. Embracing cultural perspectives can lead to more innovative and effective solutions that address both local and global environmental challenges (Bell, 2021; Ester et al., 2004.

Cultural perspectives on environmentalism provide valuable insights into how different societies relate to and manage the natural world. From Indigenous knowledge and Eastern philosophies to Western environmental movements and African stewardship, each cultural perspective offers unique contributions to environmental understanding and action. Embracing cultural diversity in environmentalism can lead to more inclusive and effective strategies, promoting a global approach that respects and incorporates diverse values and practices in the quest for a sustainable future.

9.4. Environmental Transformation in Developing vs. Developed Countries

Environmental transformation varies significantly between developing and developed countries due to differences in economic status, technological capabilities, and historical contexts. These disparities influence how nations approach environmental challenges and opportunities, shaping their strategies for sustainability, resource management, and climate action. Understanding these differences is crucial for designing effective global environmental policies and fostering equitable progress towards a sustainable future (Padilla, 2017; Batterbury et al., 1997; Rock and Angel, 2005).

9.4.1. Environmental Challenges and Priorities

9.4.1.1. Developing Countries

In developing countries, environmental transformation is often driven by the need to address pressing socio-economic issues such as poverty, inadequate infrastructure, and limited access to basic services. Environmental challenges include deforestation, land degradation, water scarcity, and air pollution, often exacerbated by rapid urbanization and industrialization. For example, deforestation in the Amazon basin, driven by agricultural expansion and logging, has significant implications for biodiversity and climate stability (Filho et al., 2015; Holdgate, 1994).

Priorities in these countries often focus on balancing economic development with environmental protection. Immediate concerns, such as improving living standards and ensuring food and water security, may take precedence over long-term environmental goals. However, there is also growing recognition of the need to incorporate sustainable practices and climate resilience into development plans.

9.4.1.2. Developed Countries

Developed countries typically face different environmental challenges, including managing the environmental impacts of high consumption levels, industrial processes, and extensive infrastructure. Issues such as greenhouse gas emissions, waste management, and energy consumption are central to environmental transformation efforts. For instance, developed countries are major contributors to global carbon emissions and must implement strategies to reduce their carbon footprint and transition to renewable energy sources (Reid, 2019).

Environmental priorities in developed countries often include advancing technological innovation, implementing comprehensive waste management systems, and promoting sustainable lifestyles. There is also a focus on addressing legacy environmental issues, such as soil and water contamination from historical industrial activities.

9.4.2. Technological Capabilities and Innovations

Technological capabilities and innovations are critical drivers of progress in addressing global environmental challenges. Breakthroughs in renewable energy, such as solar and wind power, have revolutionized the way we generate electricity, reducing reliance on fossil fuels and lowering greenhouse gas emissions. Innovations in battery storage and grid technology are essential for managing the intermittent nature of renewable energy, ensuring a stable and reliable power supply. Additionally, advancements in carbon capture and storage (CCS) hold promise for mitigating emissions from industrial processes (Corsatea, 2014; Morrison et al., 2008).

Technologies in agriculture, such as precision farming and vertical agriculture, are transforming food production, optimizing resource use, and reducing environmental impacts. Waste management technologies, like advanced recycling methods and waste-to-energy systems, are addressing pollution and resource depletion by promoting a circular economy.

However, while technological innovations offer tremendous potential, their successful implementation requires adequate funding, political support, and equitable access, particularly in developing regions where the need for sustainable solutions is most pressing.

9.4.2.1. Developing Countries

Technological capabilities in developing countries can be limited due to financial constraints and a lack of infrastructure. However, there are opportunities to leapfrog to newer, cleaner technologies that bypass traditional, more polluting practices. For example, the adoption of mobile technology for monitoring environmental conditions and managing resources has shown promise in improving environmental management and disaster response.

International support, including technology transfer and capacity-building initiatives, enables developing countries to access and implement sustainable technologies. Collaboration with international organizations, NGOs, and the private sector can facilitate the adoption of green technologies and practices.

9.4.2.2. Developed Countries

Developed countries generally have greater access to advanced technologies and research capabilities, which can drive environmental innovation. Investments in renewable energy, energy efficiency, and environmental monitoring technologies are prevalent. For instance, the development of sophisticated climate models, smart grids, and electric vehicles reflects the capacity of developed countries to lead in technological advancements for environmental sustainability.

The challenge for developed countries lies in transitioning from existing technologies and infrastructure to more sustainable alternatives, while also addressing the environmental impacts of their consumption patterns and production processes.

9.4.3. Policy Frameworks and Implementations

Policy frameworks are essential for guiding environmental action and ensuring sustainable development. They establish regulations, incentives, and goals that drive innovation and responsible resource management. Effective implementation requires a multi-level approach, involving local, national, and international cooperation. Policies like the Paris Agreement on climate change set global targets for reducing emissions, while national frameworks provide specific guidelines for sectors like energy, agriculture, and transportation. Successful implementation depends on robust monitoring, enforcement, and adaptability to evolving challenges. Additionally, inclusive participation from stakeholders ensures policies are responsive to diverse needs, promoting equity and long-term sustainability (Nilsson and Persson, 2017; Chofreh and Goni, 2017).

9.4.3.1. Developing Countries

Environmental policies in developing countries often face challenges related to implementation and enforcement due to limited resources and institutional capacity. However, there are examples of successful policy initiatives that integrate environmental and development goals. For instance, the adoption of community-based natural resource management (CBNRM) approaches has empowered local communities to manage and protect their natural resources sustainably.

International support and partnerships can enhance policy effectiveness by providing technical assistance, financial resources, and capacity-building. Initiatives such as the Green Climate Fund aim to support developing countries in implementing climate action plans and transitioning to low-carbon economies.

9.4.3.2. Developed Countries

Developed countries typically have more robust environmental policy frameworks and enforcement mechanisms. Regulations such as emissions standards, environmental impact assessments, and sustainability certifications are common. For example, the European Union's Green Deal and the United States' Clean Air Act reflect comprehensive approaches to addressing environmental challenges.

However, developed countries must continuously adapt their policies to address emerging issues and ensure that environmental regulations are effectively enforced. Additionally, there is a growing emphasis on integrating environmental considerations into economic policies and promoting corporate social responsibility.

9.4.4. Social and Economic Implications

The social and economic implications of environmental policies and technological innovations are profound, influencing livelihoods, equity, and global economies. Environmentally driven transitions, such as the shift to renewable energy, can create new jobs in green industries while phasing out traditional sectors like fossil fuels, affecting workers and communities. Policies aimed at sustainability often lead to economic restructuring, impacting local industries and global trade patterns (Williams et al., 2008; Harper, 2014).

Socially, the benefits and burdens of environmental changes are unevenly distributed, with vulnerable populations—often in developing countries—bearing the brunt of negative impacts. Inclusive strategies are crucial to ensure equitable outcomes, addressing disparities in access to resources, technology, and decision-making power. Environmental justice movements highlight the importance of fair distribution of both environmental benefits and risks.

Additionally, green initiatives can enhance public health, improve living conditions, and reduce economic disparities if implemented inclusively, ensuring that no group is disproportionately affected by environmental or technological transitions.

9.4.4.1. Developing Countries

Environmental transformation in developing countries often involves addressing socio-economic inequalities and ensuring that environmental benefits are equitably distributed. Social factors, such as access to education, healthcare, and economic opportunities, influence environmental outcomes and the ability to implement sustainable practices.

Participatory approaches that involve local communities in decision-making can enhance the effectiveness of environmental initiatives. Empowering communities to take ownership of environmental projects and incorporating traditional knowledge can lead to more sustainable and contextually appropriate solutions.

9.4.4.2. Developed Countries

In developed countries, environmental transformation can impact various social and economic dimensions, including job creation, economic competitiveness, and lifestyle changes. The transition to a green economy offers opportunities for new industries and employment, but it also requires managing the social impacts of shifting away from traditional industries.

Addressing social inequalities and ensuring a just transition for workers and communities affected by environmental policies is a key consideration. Initiatives that promote social inclusion and provide support for vulnerable populations can help mitigate the negative impacts of environmental transformations.

9.4.5. Global Cooperation and Equity

Global cooperation and equity are essential for addressing environmental challenges and achieving sustainable development. Climate change, biodiversity loss, and resource depletion are transboundary issues that require collaborative efforts between nations. Equitable cooperation ensures that both developed and developing countries share responsibilities and benefits, recognizing the historical imbalances in resource consumption and emissions (Okereke, 2008; Harris, 1996; Lange and Vogt, 2003).

For true equity, policies must prioritize support for vulnerable and marginalized communities, especially in developing nations, providing access to technology, funding, and capacity-building. International agreements, such as the Paris Agreement, emphasize fairness by balancing global cooperation with national responsibilities for an inclusive, sustainable future.

9.4.5.1. Developing Countries

Global cooperation is essential for supporting environmental transformation in developing countries. International agreements, financial mechanisms, and technology transfer initiatives play a crucial role in enabling these countries to address environmental challenges and achieve sustainable development goals.

Equity considerations are central to global environmental governance, ensuring that developing countries receive adequate support and that their voices are heard in international negotiations. The principle of common but differentiated responsibilities recognizes the varying capabilities and responsibilities of countries in addressing global environmental issues.

9.4.5.2. Developed Countries

Developed countries have a responsibility to support global environmental efforts and contribute to addressing the challenges faced by developing countries. This includes providing financial assistance, sharing technological innovations, and collaborating on research and capacity-building initiatives.

Global environmental leadership involves not only reducing domestic environmental impacts but also fostering international partnerships and promoting global equity in environmental governance.

Environmental transformation in developing and developed countries reflects diverse challenges, priorities, and capabilities. Developing countries face the dual challenge of addressing socio-economic issues while transitioning to sustainable practices, often with limited resources and technological capabilities. In contrast, developed countries grapple with managing the environmental impacts of high consumption and industrialization, leveraging advanced technologies and policy frameworks. Addressing these disparities requires global cooperation, equitable support, and context-sensitive approaches integrating local knowledge and priorities into environmental strategies. By fostering collaboration and mutual support, the global community can achieve a more sustainable and equitable future.

9.5. Global Cooperation for Environmental Transformation

Global cooperation is essential for addressing the complex and interconnected environmental challenges facing the world today. As environmental issues such as climate change, biodiversity loss, and pollution transcend national borders, collaborative efforts are crucial for achieving effective and sustainable solutions. This cooperation involves various actors, including governments, international organizations, businesses, civil society, and local communities (Brown et al., 1996; Waisová, 2015; Zinchenko et al., 2021). Understanding the mechanisms and strategies for global cooperation can enhance collective efforts toward environmental transformation and promote a more sustainable future.

9.5.1. International Agreements and Frameworks

International agreements and frameworks are crucial for coordinating global efforts to address environmental and sustainability challenges. Key agreements, like the Paris Agreement, the Convention on Biological Diversity, and the Sustainable Development Goals (SDGs), provide guidelines and commitments for countries to tackle issues such as climate change, biodiversity loss, and inequality. These frameworks promote collaboration, set targets, and establish mechanisms for monitoring progress. By fostering cooperation among nations, international agreements create a unified approach to global challenges, ensuring that policies are harmonized and equitable, with an emphasis on shared responsibilities and accountability for achieving long-term sustainability (Matz-Lück, 2012; Ioannidis et al., 2000; Stevis and Boswell, 2007).

Paris Agreement: The Paris Agreement, adopted in 2015, represents a landmark global commitment to combating climate change. It sets out ambitious goals for reducing greenhouse gas emissions, limiting global temperature rise, and enhancing climate resilience. The agreement emphasizes international cooperation and provides a framework for countries to set and achieve their climate targets through Nationally Determined Contributions (NDCs). The Paris Agreement also includes financial support and technology transfer provisions to assist developing countries' climate efforts.

Convention on Biological Diversity (CBD): Established in 1992, the CBD aims to conserve biodiversity, promote sustainable use of natural resources, and ensure fair and equitable sharing of benefits from genetic resources. The Convention fosters international collaboration through its various protocols and initiatives, including the Nagoya Protocol on Access and Benefit-sharing and the Cartagena Protocol on Biosafety. These agreements facilitate global efforts to protect ecosystems and species, support conservation programs, and promote sustainable development.

United Nations Sustainable Development Goals (SDGs): The SDGs, adopted in 2015, provide a universal framework for addressing a wide range of global challenges, including environmental sustainability. Goal 15, for example, focuses on protecting and restoring terrestrial ecosystems, while Goal 13 addresses climate action. The SDGs encourage countries to work together towards shared objectives, fostering partnerships and collaborative actions to achieve sustainable development.

9.5.2. The Role of International Organizations

International organizations play a critical role in facilitating global cooperation, setting standards, and providing support for addressing environmental, social, and economic challenges. Organizations such as the United Nations (UN), World Bank, and World Health Organization (WHO) create platforms for dialogue, coordination, and policy development among nations. They help implement international agreements, provide technical and financial assistance, and monitor progress toward global goals like the Sustainable Development Goals (SDGs). By fostering collaboration, offering expertise, and mobilizing resources, international organizations ensure that efforts to address global issues are more effective, equitable, and inclusive across regions and communities (Teixeira, 2021; Breitmeier, 1997; Tarlock, 1992).

- *United Nations Environment Programme (UNEP):* UNEP plays a central role in coordinating global environmental efforts and promoting sustainable development. It provides policy guidance, supports international environmental agreements, and facilitates the exchange of information and best practices. UNEP's initiatives include the Global Environment Outlook (GEO) reports, which assess environmental trends and provide recommendations for policy action.
- *Intergovernmental Panel on Climate Change (IPCC):* The IPCC is a leading international body for assessing climate change science and providing policy-relevant information. Its comprehensive reports synthesize the latest scientific knowledge on climate change, impacts, and mitigation strategies. The IPCC's assessments inform global climate policy and contribute to international climate negotiations.
- *International Union for Conservation of Nature (IUCN):* The IUCN focuses on nature conservation and the sustainable use of natural resources. It brings together governments, NGOs, and experts to address biodiversity conservation challenges and promote sustainable practices. The IUCN Red List of Threatened Species is a key resource for tracking species' conservation status and guiding global efforts to protect wildlife.

9.5.3. Multilateral and Bilateral Cooperation

Multilateral and bilateral cooperation are essential mechanisms for addressing global challenges, fostering international relations, and promoting sustainable development. Multilateral cooperation involves multiple countries working together through organizations like the United Nations, the World Trade Organization, or regional alliances to solve shared issues such as climate change, trade, and security. Bilateral cooperation, on the other hand, focuses on agreements between two countries, enabling more targeted collaboration on trade, technology transfer, or environmental initiatives. Both forms of cooperation are crucial for pooling resources, knowledge, and expertise, leading to stronger partnerships and more effective responses to global and regional challenges (Khan and Hou, 2021; Lo, 2009; Rotillon et al., 1996).

- *Multilateral Environmental Agreements (MEAs):* MEAs are treaties or agreements between multiple countries aimed at addressing specific environmental issues. Examples include the Montreal Protocol on Substances that Deplete the Ozone Layer, which has been successful in phasing out ozone-depleting chemicals, and the Ramsar Convention on Wetlands, which promotes the conservation and sustainable use of wetlands. Multilateral cooperation through MEAs enables countries to address transboundary environmental challenges and achieve collective goals.
- *Bilateral Partnerships:* Bilateral cooperation between countries can also play a significant role in environmental transformation. Countries may engage in joint projects, share resources and expertise, and collaborate on environmental research and conservation initiatives. For instance, bilateral agreements on forest conservation or renewable energy projects can enhance environmental outcomes and strengthen relationships between nations.

9.5.4. Engagement of Non-Governmental Organizations (NGOs) and Civil Society

NGOs and civil society play a crucial role in shaping policies, advocating for social justice, and promoting sustainable development. They act as intermediaries between governments and local communities, offering grassroots perspectives, raising awareness, and holding institutions accountable. NGOs contribute expertise in areas like environmental conservation, human rights, and poverty alleviation, often filling gaps where government efforts may be lacking. Civil society organizations foster public participation, ensuring that marginalized voices are included in decision-making processes. Their active engagement strengthens democratic governance, enhances transparency, and drives global initiatives through advocacy, education, and collaborative partnerships (Szarka, 2013; Kourula, 2010; Cooper, 2006).

- *Advocacy and Awareness*: NGOs and civil society organizations play a crucial role in raising awareness about environmental issues, advocating for policy changes, and mobilizing public support for environmental actions. Organizations such as Greenpeace, World Wildlife Fund (WWF), and Friends of the Earth work globally to address environmental challenges, promote sustainable practices, and influence policy decisions.
- *Grassroots Movements*: Grassroots movements and local advocacy groups contribute to environmental transformation by engaging communities in conservation efforts, promoting sustainable practices, and addressing local environmental issues. Community-based organizations and indigenous groups often play a vital role in protecting natural resources and ensuring environmental policies reflect local needs and knowledge.

9.5.5. Private Sector and Corporate Responsibility

The private sector plays a critical role in advancing sustainable development through corporate responsibility initiatives. Companies are increasingly expected to adopt environmentally and socially responsible practices, such as reducing carbon emissions, ensuring ethical supply chains, and supporting community development. Corporate Social Responsibility (CSR) programs promote sustainability by aligning business strategies with societal goals, including environmental protection and fair labor practices. By embracing sustainable innovations and fostering transparent governance, businesses can drive positive change while maintaining profitability. The private sector's collaboration with governments and NGOs is essential in addressing global challenges, from climate change to social inequality (Ramanathan, 2021; Rathi, 2021; Loikala et al., 2006).

- *Sustainable Business Practices*: The private sector significantly impacts environmental transformation through its practices and policies. Companies are increasingly adopting sustainable business practices, such as reducing carbon footprints, implementing circular economy principles, and investing in renewable energy. Corporate social responsibility (CSR) initiatives and sustainability reporting help companies align their operations with environmental goals and demonstrate their commitment to environmental stewardship.
- *Public-Private Partnerships*: Public-private partnerships (PPPs) can enhance environmental transformation by leveraging resources, expertise, and innovation from both sectors. Collaborations between governments, businesses, and NGOs can lead to the development of sustainable infrastructure, green technologies, and environmental conservation projects. Examples include partnerships for renewable energy development, waste management programs, and conservation initiatives.

9.5.6. Capacity Building and Knowledge Sharing

Capacity building and knowledge sharing are essential for empowering individuals, communities, and organizations to address global challenges. By enhancing skills, providing technical support, and improving access to information, capacity-building efforts foster local ownership of sustainable development initiatives. Knowledge sharing, through collaboration between governments, NGOs, academia, and the private sector, promotes the dissemination of best practices and innovative solutions. This exchange enables stakeholders to learn from diverse experiences, accelerating progress on complex issues like climate change, poverty reduction, and social equity. Effective capacity building and knowledge sharing are crucial for creating inclusive, resilient, and informed communities worldwide (Leendertse and Taylor, 2011; Mabudafhasi, 2002; Stojanovic et al., 2010).

- *Technical Assistance and Capacity Building*: Global cooperation involves providing technical assistance and capacity building to support environmental transformation in developing countries. This includes training, technology transfer, and financial support to help countries implement sustainable practices, develop environmental policies, and address capacity gaps.

- *Knowledge Sharing and Collaboration*: Sharing knowledge and best practices among countries and organizations is essential for advancing environmental transformation. International conferences, research collaborations, and knowledge platforms facilitate the exchange of information, innovations, and solutions to address global environmental challenges.

Global cooperation is a cornerstone of environmental transformation, enabling countries and organizations to address complex and transboundary environmental issues. International agreements, the role of international organizations, multilateral and bilateral partnerships, and the engagement of NGOs, the private sector, and civil society all contribute to advancing environmental sustainability. By fostering collaboration, sharing knowledge, and leveraging resources, the global community can work together to achieve meaningful and lasting environmental change, ensuring a sustainable future for future generations.

Chapter 10

Case Studies of Environmental Transformation

Case studies illustrate diverse approaches to environmental transformation across different world regions. Each example highlights unique strategies, challenges, and successes in addressing environmental issues, from large-scale reforestation and renewable energy transitions to air pollution control and water management. By examining these global efforts, we can gain valuable insights into effective environmental policies and practices, and recognize the importance of context-specific solutions and international cooperation in achieving sustainable outcomes (Mitchell et al., 2015; Dryzek et al., 2002).

10.1. Exemplary Case Studies

10.1.1. The Great Green Wall Initiative (Africa)

The Great Green Wall Initiative stands out as a major environmental transformation project in Africa, addressing desertification and land degradation in the Sahel region. The project, which began in 2007, aims to restore 100 million hectares of degraded land by creating a belt of trees and vegetation stretching across the continent. It focuses on improving soil quality, water management, and agricultural practices. The initiative has successfully restored millions of hectares, increased agricultural productivity, and enhanced local livelihoods (Goffner et al., 2019; Mbow, 2017). This case study highlights the effectiveness of large-scale reforestation efforts and community engagement in combating desertification and promoting sustainable land management.

10.1.2. Germany's Energiewende (Europe)

Germany's Energiewende, or "energy transition," represents a comprehensive strategy to shift from fossil fuels and nuclear energy towards renewable energy sources. Initiated in the early 2000s, the policy aims to reduce greenhouse gas emissions, phase out nuclear power, and increase the share of renewables in the energy mix. The Energiewende has led to significant investments in wind and solar energy, technological innovations, and a substantial reduction in carbon emissions (Beveridge and Kern, 2013; Buchan, 2012; Lauber and Jacobsson, 2015). The case study demonstrates how ambitious national policies can drive substantial environmental transformation, promote sustainable energy practices, and create economic opportunities in the renewable energy sector.

10.1.3. Costa Rica's Conservation and Renewable Energy Efforts (Central America)

Costa Rica is a leading example of environmental transformation in the realm of conservation and renewable energy. By the early 2000s, Costa Rica had committed to a bold strategy of forest conservation, reforestation, and sustainable land management. The country has successfully reversed deforestation trends, restored significant areas of rainforest, and protected biodiversity. Additionally, Costa Rica generates over 99% of its electricity from renewable sources, primarily hydroelectric, wind, and solar power (Dolezal et al., 2013; Green and Barborak, 1987; Teske et al., 2020). This case study illustrates the potential for small countries to achieve significant environmental and climate goals through comprehensive conservation efforts and a commitment to renewable energy.

10.1.4. China's Air Pollution Control Measures (Asia)

China's rapid industrialization has led to severe air pollution challenges, but the country has undertaken significant measures to address this issue. Initiatives include the implementation of stricter emissions standards, the promotion of clean energy technologies, and large-scale investments in public transportation. The "Air Pollution Prevention and Control Action Plan" and subsequent policies have led to improvements in air quality, reduced particulate matter levels, and increased public health outcomes (Jin et al., 2016; Hu et al., 2010). This case study demonstrates the challenges and successes of implementing environmental policies in rapidly industrializing economies and highlights the importance of targeted interventions and regulatory measures.

10.1.5. Australia's Murray-Darling Basin Plan (Oceania)

The Murray-Darling Basin Plan is a comprehensive strategy aimed at addressing water management issues in Australia's largest river system. The plan, developed in response to severe droughts and ecological degradation, focuses on sustainable water allocation, improving water quality, and restoring riverine ecosystems. The initiative has led to improvements in water availability, ecological health, and agricultural practices (Prosser et al., 2021; Stewardson et al., 2021). This case study underscores the importance of integrated water management approaches and the need for collaboration between government agencies, local communities, and stakeholders in addressing complex environmental challenges.

10.1.6. India's Swachh Bharat Mission (South Asia)

India's Swachh Bharat Mission (Clean India Mission), launched in 2014, aims to improve sanitation and hygiene nationwide. The mission focuses on increasing access to clean and safe sanitation facilities, promoting hygiene practices, and reducing open defecation. The initiative has led to constructing millions of toilets, improved waste management, and enhanced public health (Dandabathula et al., 2019; Jangra et al., 2016). This case study illustrates how targeted national programs can address critical environmental and health issues and highlights the role of government leadership and community participation in achieving transformative outcomes.

10.1.7. Nepal's Community Forestry Program (CFP)

CFP is one of the most successful and innovative examples of environmentalism and participatory natural resource management worldwide. Established in the late 1970s, this initiative empowers local communities to manage, conserve, and utilize their forests, aligning environmental sustainability with socio-economic development.

The core principle of Nepal's CFP is decentralization and community ownership. Under this program, the government transfers the management rights of national forests to local communities, forming Community Forest User Groups (CFUGs). These groups protect, sustainably harvest, and regenerate forest resources. Today, more than 22,000 CFUGs manage approximately 40% of Nepal's forested areas, benefiting millions of households.

One of the CFP's key best practices is its participatory approach. CFUGs are inclusive and democratic, allowing members, including women and marginalized groups, to participate actively in decision-making processes. This promotes equity, ensuring that the benefits of forest resources are shared fairly among all members and strengthening social cohesion within communities.

Another successful element is the focus on sustainability. CFUGs must develop operational plans that outline the sustainable use of forest resources, including the regulated collection of firewood, fodder, and timber. These plans also emphasize forest conservation through reforestation and soil erosion control, ensuring that forests remain viable for future generations.

Furthermore, the CFP has demonstrated environmental and economic success. Forest cover in Nepal has increased due to effective community management, reversing decades of deforestation. Economically, the program has provided communities with income by selling forest products and non-timber resources, improving livelihoods and reducing poverty.

The CFP's adaptability is also a best practice. It has evolved to address challenges such as climate change, forest fires, and biodiversity loss, ensuring that Nepal's forests thrive (Luintel et al., 2017, 2018; Upreti, 2000). In sum, Nepal's Community Forestry Program is a model of sustainable environmentalism, balancing conservation with community development.

10.2. Renewable Energy Transitions in Europe and Asia

Renewable energy transitions in Europe and Asia reshape global energy landscapes and promote sustainability. Europe has made significant wind and solar energy strides, driven by ambitious climate policies and technological innovations. Countries like Germany and Denmark lead in renewable installations, reducing reliance on fossil fuels. In Asia, nations such as China and India are rapidly expanding solar and wind capacities to meet growing energy demands while addressing environmental concerns (Wang et al., 2018; Radovanović et al., 2021). These transitions enhance energy security, create economic opportunities, foster green jobs, and contribute to global efforts in combating climate change and promoting sustainable development.

10.2.1. Renewable Energy Transitions in Europe

Europe has been at the forefront of renewable energy transitions, driven by ambitious climate policies, technological advancements, and a commitment to sustainability. Several key aspects characterize Europe's approach to renewable energy (Allen et al., 2021; Ćetković and Buzogány, 2016).

- *Policy Frameworks and Goals*: European countries have established robust policy frameworks to promote renewable energy. The European Union (EU) has set ambitious targets through its Renewable Energy Directive, aiming to achieve a 32% share of renewable energy in the EU's energy mix by 2030. Countries like Germany, Denmark, and Spain have implemented national policies and incentives to support renewable energy development, including feed-in tariffs, subsidies, and tax benefits.
- *Technological Innovation*: Europe has been a leader in renewable energy technology, particularly in wind and solar power. Denmark is renowned for its wind energy achievements, with wind turbines contributing a significant portion of the country's electricity generation. Germany's Energiewende has driven significant advancements in solar photovoltaic (PV) technology and energy storage solutions. European countries are also investing in emerging technologies such as green hydrogen and offshore wind farms.
- *Infrastructure and Integration*: The integration of renewable energy into the grid has been a priority, with significant investments in smart grid technologies and cross-border electricity networks. The EU's Clean Energy for All Europeans package includes measures to enhance grid connectivity, support energy storage, and enable cross-border energy trading. These efforts facilitate the efficient use of renewable energy and enhance energy security across the region.
- *Economic and Social Impacts*: The renewable energy transition in Europe has created job opportunities, fostered economic growth, and contributed to energy independence. The growth of the renewable energy sector has also increased public awareness and support for sustainability initiatives. However, challenges remain, such as managing the intermittency of renewable sources and ensuring equitable energy transitions for all regions and communities.

10.2.2. Renewable Energy Transitions in Asia

Asia is experiencing rapid renewable energy transitions driven by growing energy demands, environmental concerns, and economic development (Marquardt, 2016; Vakulchuk et al., 2023).

- *Diverse Strategies and Policies*: Asian countries have adopted various strategies to promote renewable energy, reflecting their unique resources and priorities. As the world's largest emitter and energy consumer, China has made substantial investments in solar and wind energy. The country's 13th Five-Year Plan includes ambitious targets for renewable energy capacity and green technology development. India has also set aggressive goals under its National Solar Mission to achieve 100 GW of solar power capacity by 2022.
- *Technological Advancements:* Asia is making significant strides in renewable energy technology. China is a global leader in solar PV manufacturing and deployment, with large-scale solar farms and innovations in solar technology. Japan has focused on advancing energy storage solutions and offshore wind energy. South Korea invests in next-generation renewable technologies, including floating solar and hydrogen energy.
- *Challenges and Opportunities*: Asia faces challenges in its renewable energy transitions, including balancing energy demands with environmental goals, managing grid integration, and addressing financing barriers. However, the region also presents growth opportunities, particularly in harnessing abundant solar and wind resources. Collaborative efforts, such as regional partnerships and investments in green technology, are crucial for advancing renewable energy across Asia.
- *Economic and Social Impacts*: The transition to renewable energy in Asia is driving economic development, creating jobs, and improving energy access. Initiatives such as the International Solar Alliance (ISA), which promotes solar energy deployment in tropical countries, reflect the region's commitment to global cooperation. The renewable energy sector is also contributing to energy security and reducing reliance on fossil fuels, while efforts to address environmental impacts and ensure inclusive growth continue.

Renewable energy transitions in Europe and Asia illustrate diverse approaches and strategies for advancing sustainability and addressing climate challenges. Europe's leadership in renewable technology, supportive policies, and infrastructure development contrast with Asia's rapid growth, technological advancements, and unique challenges. Both regions demonstrate the potential for renewable energy to drive economic growth, enhance energy security, and contribute to global climate goals. Collaboration and knowledge-sharing between Europe and Asia can further accelerate progress and support a sustainable global energy future.

10.3. Community-Based Conservation in Latin America

Community-based conservation in Latin America represents a transformative approach to environmental management, emphasizing the active participation of local communities in the stewardship of natural resources. This model recognizes that local people, who are often deeply connected to their environments, are essential to achieving sustainable conservation outcomes. Several key aspects highlight the successes, challenges, and impacts of community-based conservation in the region (Ruiz-Mallén et al., 2015; del Mar Delgado-Serrano et al., 2017).

10.3.1. Indigenous Land Stewardship

Indigenous communities in Latin America have long practiced sustainable land management, guided by traditional ecological knowledge and cultural values. For example, the Amazonian communities in countries like Brazil, Ecuador, and Peru have developed intricate systems for managing forests, waterways, and wildlife. Initiatives such as the Yasuni-ITT Initiative in Ecuador, which sought to leave oil reserves untapped in exchange for international funding, reflect the growing recognition of indigenous stewardship (Sangha, 2020; Schang et al., 2020). The integration of indigenous knowledge with modern conservation practices often leads to more effective and culturally sensitive environmental management.

10.3.2. Community-Managed Protected Areas

Community-managed protected areas are a significant component of conservation efforts in Latin America. The development of these areas often involves formal agreements between local communities and national or international conservation organizations. One notable example is the Sarayaku Indigenous Territory in Ecuador, where local communities have successfully managed their land to protect biodiversity and resist oil exploration. Similarly, the Reserva de la Biosfera Maya in Guatemala represents a collaborative effort between local communities, NGOs, and the government to conserve tropical forests and wildlife (Porter-Bolland et al., 2012; Rao et al., 2016).

10.3.3. Eco-Tourism and Sustainable Livelihoods

Eco-tourism has become a valuable tool for community-based conservation in Latin America, providing economic incentives for environmental protection. In countries like Costa Rica and Peru, eco-tourism ventures managed by local communities offer opportunities for sustainable income while preserving natural habitats. For instance, the Tambopata National Reserve in Peru has developed eco-tourism initiatives that involve local indigenous groups in guiding and managing tourism activities (Choudhury, 2019; Bhattacharya, 2005; Bhandari, 2014; Sapiński et al., 2022). These ventures contribute to conservation funding, promote cultural heritage, and reduce pressures on natural resources.

10.3.4. Challenges and Conflicts

Despite its successes, community-based conservation in Latin America faces several challenges. Conflicts may arise between conservation goals and local economic needs, such as when communities depend on resource extraction for livelihoods. Additionally, external pressures, including deforestation, mining, and agricultural expansion, can undermine conservation efforts (Novelli et al., 2012; Raj and Griffin, 2017; Al Haija, 2011). Ensuring that conservation strategies are equitable and respect local rights is crucial for addressing these challenges and achieving long-term sustainability.

10.3.5. Policy and Institutional Support

Effective community-based conservation often requires supportive policies and institutional frameworks. Latin American countries have made strides in recognizing the rights of indigenous peoples and community land management through legislation and agreements. For example, Brazil's Forest Code and Colombia's National Program for the Management of Protected Areas include provisions for community involvement in conservation (Dredge, 2017; Bousset et al., 2007). However, implementation gaps and bureaucratic hurdles can limit the effectiveness of these policies.

10.3.6. Examples of Successful Initiatives

Several successful community-based conservation initiatives in Latin America illustrate the potential of this approach. The Community Forest Management project in Bolivia empowers local communities to manage and benefit from forest resources while conserving biodiversity. Similarly, the Programa de Conservación de Tortugas Marinas (Marine Turtle Conservation Program) in Nicaragua involves local communities in protecting nesting sites and monitoring turtle populations, contributing to the recovery of endangered species.

Community-based conservation in Latin America demonstrates the power of local knowledge, participation, and management in achieving sustainable environmental outcomes. By integrating indigenous practices, promoting eco-tourism, and fostering collaborative management of protected areas, this approach offers valuable lessons for conservation efforts globally (Simpson, 2008; Jewell et al., 2004; Kayat and Zainuddin, 2016). Addressing challenges and enhancing policy support are essential for maximizing the benefits of community-based conservation and ensuring its long-term success in the region.

10.4. Urban Sustainability Initiatives in North America

Urban sustainability initiatives in North America are transforming cities into more environmentally friendly, economically viable, and socially inclusive spaces. As urban areas grapple with challenges such as climate change, resource depletion, and social inequality, a range of innovative approaches and strategies are being employed to promote sustainability (National Academies of Sciences, Policy, Global Affairs, Technology for Sustainability Program, Committee on Pathways to Urban Sustainability, and Opportunities, 2016; Fitzgerald, 2010; Alberti, 1996).

10.4.1. Green Building Standards and Certifications

Green building standards have become a cornerstone of urban sustainability in North America. Programs such as Leadership in Energy and Environmental Design (LEED) and the Green Building Initiative's Green Globes provide frameworks for designing and constructing environmentally responsible buildings. Cities like Vancouver and New York have implemented stringent green building codes and incentivized sustainable construction practices. Vancouver's Green Building Policy requires new city-owned buildings to achieve LEED Gold certification or higher, significantly reducing their environmental impact (Vierra, 2016; Mell, 2009).

10.4.2. Renewable Energy Integration

Urban centers are increasingly integrating renewable energy sources to reduce their reliance on fossil fuels and lower greenhouse gas emissions. San Francisco, for instance, has committed to running its municipal operations on 100% renewable energy. Similarly, Toronto's Renewable Energy Standard requires all new buildings to include renewable energy systems, such as solar panels (Abdmouleh et al., 2015; Oró et al., 2015). Urban areas also invest in community solar projects and renewable energy incentives to encourage residential and commercial adoption of green energy technologies.

10.4.3. Sustainable Transportation Systems

Sustainable transportation is a critical component of urban sustainability. Many North American cities are investing in public transportation infrastructure, cycling networks, and pedestrian-friendly urban design to reduce traffic congestion and emissions. Portland, Oregon, is known for its extensive network of bike lanes and commitment to increasing cycling infrastructure. Similarly, the implementation of electric bus fleets in cities like

Los Angeles and New York reflects efforts to transition to cleaner public transit options (Rajak et al., 2016; Richardson, 1999; Awasthi et al., 2011).

10.4.4. Urban Green Spaces and Parks

The creation and maintenance of urban green spaces and parks contribute significantly to environmental and social sustainability. New York City's High Line, an elevated park built on a former railway track, exemplifies innovative use of urban space to provide green areas, improve air quality, and enhance community well-being. Chicago's Millennium Park and the expansion of the Toronto waterfront also highlight efforts to integrate nature into urban environments, offering recreational opportunities and improving urban resilience (Fors et al., 2015; Halecki et al., 2023; Özgüner, 2011).

10.4.5. Waste Management and a Circular Economy

North American cities are increasingly adopting waste management practices that emphasize recycling, composting, and the circular economy. Cities like Toronto and San Francisco have implemented comprehensive recycling and composting programs, achieving high diversion rates from landfills. San Francisco's Zero Waste by 2020 goal focuses on reducing waste generation and promoting material recovery (Hidalgo et al., 2019; Luttenberger, 2020; Tomić and Schneider, 2020). Additionally, initiatives such as composting programs and waste-to-energy facilities reflect a commitment to reducing environmental impact and promoting sustainable resource use.

10.4.6. Climate Action Plans

Many North American cities have developed climate action plans to address the impacts of climate change and build resilience. For example, Los Angeles has the Sustainable City pLAn, which outlines strategies for reducing greenhouse gas emissions, improving air quality, and enhancing climate resilience. Similarly, Montreal's Climate Plan aims to achieve carbon neutrality by 2050, focusing on energy efficiency, renewable energy, and sustainable transportation (Stone et al., 2012; Bassett and Shandas, 2010).

These plans set ambitious targets and provide a roadmap for integrating sustainability into urban planning and development.

10.4.7. Water Conservation and Management

Water conservation and management are critical aspects of urban sustainability, particularly in regions facing water scarcity. Cities like Phoenix and Las Vegas are implementing water-efficient technologies and practices to manage their water resources sustainably. Initiatives include the use of low-water landscaping, water recycling programs, and public education campaigns on water conservation (Blanco-Canqui et al., 2008; Maggioni, 2015). Additionally, integrating green infrastructure, such as rain gardens and permeable pavements, helps manage stormwater and reduce runoff.

10.4.8. Community Engagement and Equity

Urban sustainability initiatives increasingly focus on community engagement and social equity. Programs that involve local residents in decision-making processes and ensure equitable access to sustainability benefits are vital for creating inclusive and resilient cities. Initiatives such as community gardens, local sustainability workshops, and participatory planning processes in cities like Seattle and Boston aim to empower residents and address environmental justice concerns (Telles, 2019; Boston et al., 2015; Grumbach et al., 2017).

Urban sustainability initiatives in North America reflect a comprehensive approach to addressing cities' environmental, economic, and social challenges. Cities are making strides toward more sustainable futures by adopting green building standards, renewable energy integration, sustainable transportation, and effective waste management. The emphasis on community engagement and equity ensures that these initiatives benefit all residents and contribute to more resilient urban environments. By continuing to innovate and collaborate, North American cities are setting examples for sustainable urban development globally.

10.5. Sustainable Agricultural Practices in Africa

Sustainable agricultural practices in Africa are vital for enhancing food security, preserving natural resources, and promoting economic development. With Africa's diverse climates and ecosystems, sustainable agriculture addresses unique challenges while leveraging local knowledge and innovations (Behnassi et al., 2011; Nkomoki et al., 2018).

10.5.1. Agroforestry

Agroforestry integrates trees and shrubs into agricultural systems, providing multiple benefits such as improved soil fertility, enhanced biodiversity, and increased resilience to climate change. In countries like Kenya and Ethiopia, farmers practice agroforestry by planting trees alongside crops and livestock. These trees help prevent soil erosion, improve water retention, and provide shade for crops and animals (Nair et al., 2017; Maroyi, 2009). Additionally, tree products such as fruits, nuts, and timber offer additional income sources for farmers.

10.5.2. Conservation Agriculture

Conservation agriculture focuses on maintaining soil health and reducing erosion through minimal soil disturbance, cover cropping, and crop rotation. This practice is particularly relevant in regions like Southern Africa, where soil degradation is a significant concern. Techniques such as no-till farming and cover crops like legumes enhance soil structure, increase organic matter, and promote water retention. Conservation agriculture helps improve yields, reduce input costs, and mitigate the impacts of climate variability (Dumanski et al., 2006; Fonteyne et al., 2022).

10.5.3. Climate-Smart Agriculture

Climate-smart agriculture (CSA) aims to increase productivity while reducing greenhouse gas emissions and enhancing resilience to climate change. CSA practices include improving crop varieties, adopting efficient irrigation techniques, and using organic fertilizers. In Mali and Ghana, CSA initiatives support farmers in adapting to changing weather patterns and optimizing resource use. For example, using drought-resistant crop varieties and precision irrigation systems helps mitigate the effects of climate extremes and ensure stable food production (Dev et al., 2022; Udawatta et al., 2021).

10.5.4. Integrated Pest Management (IPM)

Integrated Pest Management (IPM) is a holistic approach to managing pests and diseases using biological, cultural, and chemical methods. IPM reduces the reliance on chemical pesticides, which can have harmful environmental and health impacts. Farmers in countries like Nigeria and Uganda use IPM strategies such as introducing natural predators, practicing crop rotation, and monitoring pest populations to manage pests more sustainably (Ehler, 2006; Elliott et al., 1995). IPM helps protect crops, reduce pesticide use, and promote ecological balance.

10.5.5. Sustainable Livestock Management

Sustainable livestock management practices aim to enhance the productivity and health of livestock while minimizing environmental impacts. In regions like East Africa, practices such as rotational grazing, improved fodder management, and water-efficient livestock systems are being implemented. Rotational grazing allows pastures to recover and improves soil health, while improved fodder management ensures livestock receive balanced nutrition (Norton, 1998; Kemp and Michalk, 2007). Sustainable livestock practices help reduce overgrazing, improve animal welfare, and increase farm productivity.

10.5.6. Water Harvesting and Management

Effective water management is crucial for sustainable agriculture, particularly in drought-prone regions. Techniques such as rainwater harvesting, soil moisture conservation, and drip irrigation are used to optimize water use. In countries like Ethiopia and Kenya, farmers construct rainwater harvesting ponds and use drip irrigation systems to provide consistent crop water supply (Sivanappan, 2006; Datta, 2019; Oweis et al., 2001). These practices help increase crop yields, reduce water wastage, and enhance resilience to dry periods.

10.5.7. Organic Farming

Organic farming emphasizes using natural inputs and processes to maintain soil fertility, control pests, and produce healthy food. In Africa, organic farming is gaining popularity as a way to reduce dependence on synthetic chemicals and promote environmentally friendly practices. Countries like Uganda and Tanzania have seen growth in organic farming, with farmers using compost, green manures, and natural pest control methods (Lampkin et al., 2000; Reddy, 2010; Seufert et al., 2017). Organic farming helps improve soil health, enhance biodiversity, and provide access to niche markets.

10.5.8. Community-Based Approaches

Community-based approaches to agriculture involve local farmers and stakeholders in decision-making processes and promote collaborative efforts. Initiatives such as farmer cooperatives, community seed banks, and shared knowledge networks empower communities to manage their resources sustainably. For example, the Participatory Community-Based Development Project in Burkina Faso involves local farmers in planning and implementing agricultural activities. These approaches build local capacity, enhance collective action, and support sustainable development.

Sustainable agricultural practices in Africa are essential for addressing food security, environmental degradation, and economic challenges. By adopting practices such as agroforestry, conservation agriculture, climate-smart agriculture, and integrated pest management, African farmers are enhancing productivity and resilience while protecting natural resources. The emphasis on community-based approaches and innovative techniques ensures that these practices are adapted to local conditions and contribute to long-term sustainability (Schilderman, 2004; Deakin, 2009; Carr and Halvorsen, 2001). As Africa faces climate and resource challenges, sustainable agriculture will be critical in building a resilient and prosperous future for its communities.

10.6. Technology-Driven Solutions in the Middle East

Technology-driven solutions are playing a pivotal role in addressing various challenges in the Middle East, a region characterized by its diverse landscapes, rapidly growing populations, and economic aspirations. From water scarcity and energy demands to urbanization and healthcare, technology is being leveraged to foster sustainable development and enhance the quality of life (Jalbout and Farah, 2016; Kheira, 2021).

10.6.1. Water Management and Desalination

Water scarcity is a critical issue in the Middle East, prompting advancements in water management and desalination technologies. Countries like Saudi Arabia, the UAE, and Israel have invested heavily in desalination plants to convert seawater into potable water. For example, Saudi Arabia operates one of the world's largest desalination facilities, the Ras Al Khair plant, which utilizes reverse osmosis and multi-effect distillation technologies.

In addition to desalination, water conservation and efficient usage technologies are being implemented. Smart irrigation systems, which use sensors and data analytics to optimize water application for agriculture, are becoming increasingly common. The UAE's drone technology for monitoring irrigation practices exemplifies how technology can enhance water management (Adibfar, 2013).

10.6.2. Renewable Energy Technologies

The Middle East, traditionally reliant on fossil fuels, is making significant strides in renewable energy. Given the region's high solar radiation levels, solar power is a major focus. The UAE's Masdar City, a planned sustainable urban development in Abu Dhabi, features a large solar park contributing to the city's goal of becoming a zero-carbon environment. Similarly, Saudi Arabia's King Abdulaziz City for Science and Technology invests in solar and wind energy projects to diversify its energy portfolio.

Additionally, Bahrain and Qatar are exploring the potential of wind energy, with projects like the Bahrain Wind Farm and Qatar's interest in offshore wind technology (Krupa et al., 2019). These initiatives reflect the region's commitment to transitioning to sustainable energy sources.

10.6.3. Smart Cities and Urban Development

The concept of smart cities is gaining traction in the Middle East, with technology being integrated into urban planning and management to improve efficiency and livability. Cities like Dubai and Riyadh are leading the way in this regard. Dubai's Smart City initiative includes projects such as smart traffic management systems, energy-efficient buildings, and digital platforms for city services.

Riyadh's King Abdullah Financial District development incorporates smart technologies for infrastructure management, energy efficiency, and sustainability (El-Kholei and Al Jayyousi, 2019; Ali et al., 2023). These smart city projects aim to enhance urban living by leveraging data and technology to optimize resources and services.

10.6.4. Health Tech and Telemedicine

The healthcare sector in the Middle East is increasingly adopting digital health technologies to improve patient care and streamline services. Telemedicine platforms, such as those developed by the UAE's Health Authority and Saudi Arabia's Ministry of Health, provide remote consultation and treatment options, particularly valuable in remote and underserved areas.

Additionally, artificial intelligence (AI) is advancing in diagnostics and treatment planning. For example, AI-driven imaging tools are being used to enhance the accuracy of medical diagnoses and support personalized treatment plans. The Qatar Biomedical Research Institute also explores AI applications in genomics and drug discovery (Al-Shorbaji and Alhuwail, 2022; Mansour, 2022).

10.6.5. Educational Technology and E-Learning

Educational technology is transforming learning experiences in the Middle East, focusing on digital tools and e-learning platforms. The UAE's initiatives, such as the Smart Learning Program, aim to integrate technology into classrooms to enhance educational outcomes. Virtual and augmented reality (VR/AR) create immersive learning environments, making complex subjects more engaging and accessible.

Saudi Arabia's Vision 2030 also emphasizes the role of technology in education, with investments in digital learning platforms and online courses to improve educational quality and access (Jalbout and Farah, 2016; El-Kogali and Krafft, 2019). The rise of online education providers and educational apps reflects a growing emphasis on technology-driven learning solutions.

10.6.6. Fintech and Digital Payment Systems

The fintech sector is rapidly growing in the Middle East, with digital payment systems, blockchain technology, and financial services innovations. Countries like the UAE and Bahrain are leading the way in fintech adoption, with initiatives like the Dubai International Financial Centre's FinTech Hive, which supports startups and innovation in financial technology.

Blockchain technology is also being explored for applications in banking, supply chain management, and security. For example, Dubai's Blockchain Strategy aims to transform government transactions and services using blockchain technology, enhancing transparency and efficiency (Kheira, 2021; Sujee and Solanki, 2024).

10.6.7. Agriculture Technology

Agriculture in the Middle East is benefiting from technology-driven solutions to improve productivity and sustainability. Precision agriculture technologies, such as remote sensing and data analytics, optimize irrigation, manage soil health, and increase crop yields. The UAE's International Center for Biosaline Agriculture focuses on research and development of saline-tolerant crops and advanced irrigation techniques to address the challenges of arid environments.

Technology-driven solutions in the Middle East address arious critical challenges and drive progress in various sectors. Technology is transforming the region's future, from advanced water management and renewable energy to smart cities and healthcare innovations. By continuing to invest in and leverage technological advancements, the Middle East is positioning itself as a sustainable development and innovation leader (Mati and Penning de Vries, 2005; Lockie et al., 2020).

Chapter 11

The Future of Environmentalism: Towards a Sustainable Planet

The future of environmentalism is increasingly shaped by a blend of innovative technologies, evolving societal values, and a heightened awareness of the urgent need for sustainable practices. As we look toward the future, several key trends and factors are poised to define the trajectory of environmentalism and drive progress toward a sustainable planet (Dawson, 2024; Tavanti and Sfeir-Younis, 2024).

11.1. The Future of Environmentalism

The future of environmentalism is shaped by growing global awareness of the urgent need for sustainable practices, technological innovations, and inclusive policies. As environmental issues such as climate change, biodiversity loss, and pollution continue to escalate, environmentalism is evolving from a niche movement to a mainstream priority for governments, corporations, and civil society (Dawson, 2024; Tavanti and Sfeir-Younis, 2024).

One key aspect of environmentalism's future is technology's increasing role. Innovations like renewable energy, smart cities, and sustainable agriculture will be pivotal in reducing environmental impacts and promoting efficiency. However, the challenge will be to ensure these technologies are accessible and equitably distributed, particularly in developing regions.

Environmental justice and intersectionality will also be central to the movement. Future environmentalism must address the disproportionate impacts of environmental degradation on marginalized communities, integrating social equity into environmental policies and practices. This approach emphasizes that the fight for environmental sustainability cannot be separated from broader human rights and social justice struggles (Castree, 2006; McGrail, 2011; Wapner, 2013).

Moreover, the future of environmentalism will depend on stronger global cooperation. International agreements, corporate accountability, and grassroots activism will continue to be vital in driving systemic changes, ensuring that environmental policies and actions are both effective and inclusive, fostering a sustainable and just future for al (O'Riordan, 2004; McGrail, 2011; O'Riordan, 1991).

11.1.1. The Integration of Technology and Environmental Science

One of the most significant shifts in the future of environmentalism is the growing integration of technology with environmental science. Advanced technologies such as artificial intelligence (AI), big data, and the Internet of Things (IoT) are transforming how we monitor and manage environmental resources. AI and machine learning algorithms can analyze vast amounts of data to predict environmental changes, optimize resource use, and identify patterns that traditional methods might miss. For example, satellite technology combined with AI is increasingly used for tracking deforestation, monitoring wildlife populations, and assessing the impacts of climate change in real-time (Rennings et al., 2003; Argent, 2004; ChanLin, 2008). These technological advancements offer unprecedented opportunities for enhancing environmental protection and sustainability.

11.1.2. Emphasis on Circular Economy

The circular economy concept is gaining momentum as a crucial element of future environmentalism. Unlike the traditional linear model of "take-make-dispose," a circular economy focuses on designing products and systems that minimize waste, maximize resource efficiency, and promote recycling and reuse. This means developing products with longer lifespans, implementing closed-loop production processes, and creating systems for recovering and repurposing materials. Businesses and governments increasingly adopt circular economy principles to reduce environmental impact and create more sustainable economic models (Arruda et al., 2021; Kirchherr et al., 2017; Moustakas and Loizidou, 2023). This shift represents a fundamental change in how we approach consumption and waste, moving towards a more regenerative and sustainable way of living.

11.1.3. Greater Focus on Climate Adaptation and Resilience

As the impacts of climate change become more pronounced, there is an increasing emphasis on climate adaptation and resilience. While mitigation efforts to reduce greenhouse gas emissions remain crucial, adapting to the inevitable effects of climate change is equally important. Future environmentalism will involve strategies to build resilience in both natural and human systems. This includes developing infrastructure that can withstand extreme weather events, implementing community-based adaptation programs, and protecting and restoring ecosystems that provide vital services such as flood control and water purification (Nelson, 2011; Abeling et al., 2018; Davoudi et al., 2013). For instance, creating climate-resilient urban landscapes that incorporate green infrastructure and natural solutions will be essential for mitigating the impacts of climate change on cities and vulnerable communities.

11.1.4. Strengthening Global Collaboration and Governance

Addressing global environmental challenges requires enhanced international cooperation and governance. The future of environmentalism will be characterized by more robust global frameworks for managing shared resources and addressing transboundary issues. International agreements such as the Paris Agreement on Climate Change and the Convention on Biological Diversity provide a foundation for collaborative efforts. However, more ambitious and coordinated actions will be needed. Strengthening global environmental governance involves not only enhancing existing treaties and organizations but also fostering new forms of collaboration that include diverse stakeholders, such as local communities, indigenous peoples, and the private sector (Jakab et al., 2021; Thakur and Van Langenhove, 2007; Asaduzzaman et al., 2016). Collaborative efforts can drive more effective solutions and ensure that environmental policies are implemented equitably across different regions.

11.1.5. Inclusivity and Equity in Environmentalism

The future of environmentalism will increasingly prioritize inclusivity and equity, recognizing that environmental issues are deeply intertwined with social justice. Ensuring that all communities, particularly marginalized and disadvantaged groups, have a voice in environmental decision-making is essential for achieving sustainable outcomes. This includes addressing environmental justice issues such as exposure to pollution, access to clean resources, and the impacts of climate change. Environmental equity emphasizes that the benefits and burdens of environmental policies should be distributed fairly and that marginalized communities should be empowered to participate in and benefit from environmental initiatives (Teelucksingh, 2019; Engle et al., 2024; Lu et al., 2024). Embracing equity in environmentalism will help build more resilient and just societies.

11.1.6. Transformation of Consumer Behavior and Lifestyle

Future environmentalism will also involve a transformation in consumer behavior and lifestyle choices. As awareness of environmental issues grows, there is a shift towards more sustainable consumption patterns and lifestyle choices. This includes increased adoption of renewable energy, reduction in single-use plastics, and support for sustainable and ethical products. The rise of the sharing economy, where resources are shared rather than owned, and the growth of plant-based diets reflect changing attitudes towards sustainability. Encouraging and facilitating these changes in consumer behavior will be critical for reducing environmental impacts and promoting a more sustainable future (Yegina et al., 2020; Schreurs et al., 2012; Febiyanti and Nadlifatin, 2024). The future of environmentalism is characterized by a dynamic interplay of technology, policy, and societal values, all aimed at achieving a sustainable planet. The integration of advanced technologies, adoption of circular economy principles, focus on climate adaptation, strengthening of global collaboration, and commitment to equity are all pivotal elements shaping the path forward. As we navigate these changes, fostering a holistic and inclusive approach to environmentalism will be essential for addressing our complex challenges and ensuring a sustainable and equitable future for all.

11.1.7. Foresight and Future Scenarios for Environmental Transformation

Foresight and future scenarios are crucial in shaping the strategies and policies needed for environmental transformation. By anticipating potential developments and exploring various pathways, stakeholders can better prepare for uncertainties and make informed decisions to guide sustainable progress (Hebinck et al., 2018; Larsen and Hojer, 2007; Hussain et al., 2017).

11.2. Foresight and Future Scenarios

Foresight in environmentalism involves anticipating future challenges and opportunities to shape sustainable policies and actions. Future scenarios of environmentalism could include accelerated climate change impacts, increasing biodiversity loss, and heightened resource scarcity, demanding urgent responses. Conversely, advancements in renewable energy, circular economies, and eco-friendly technologies could lead to more sustainable lifestyles and economic models. Collaborative governance, global partnerships, and equitable resource distribution will be crucial in navigating these scenarios (Hebinck et al., 2018). Integrating social justice, technological innovation, and environmental resilience will define the future landscape of environmentalism, ensuring sustainability in a rapidly changing world.

11.2.1. Scenario Planning and Modeling

Scenario planning and modeling are essential for envisioning different future pathways and their potential environmental impacts. These methods involve creating a range of plausible scenarios based on current trends, uncertainties, and possible disruptions. For example, environmental organizations and policymakers use scenario modeling to project the impacts of climate change under different greenhouse gas emission trajectories. Such models help understand potential future conditions, from worst-case scenarios with severe climate impacts to more optimistic scenarios with significant mitigation efforts (Peterson et al., 2003). Scenario planning allows for the development of adaptive strategies that can be adjusted as conditions evolve, ensuring that policies and actions remain relevant and effective.

11.2.2. Emerging Technologies and Innovations

Emerging technologies and innovations heavily influence the future of environmental transformation. Advances in renewable energy, biotechnology, and information technology will likely drive significant changes in how we address environmental challenges. For instance, breakthroughs in solar and wind energy technologies could lead to more efficient and cost-effective renewable energy systems, accelerating the transition from fossil fuels. Similarly, innovations in biotechnology may offer new solutions for environmental restoration, such as genetically engineered plants that can thrive in degraded soils or absorb pollutants (Srinivasan, 2008). Staying abreast of technological developments and incorporating them into future scenarios can help anticipate their potential impacts and opportunities for environmental transformation.

11.2.3. Socio-Economic Shifts and Policy Responses

Foresight also involves considering socio-economic shifts and their implications for environmental policy. Changes in demographics, economic structures, and societal values can significantly influence environmental outcomes. For example, increasing urbanization and population growth may drive higher demand for resources and energy, necessitating innovative solutions for sustainable urban development. Additionally, shifts in consumer behavior towards more sustainable practices can impact production and consumption patterns. Policymakers must anticipate these socio-economic changes and design policies that address current and future needs (Van Ginkel, 2020). This includes developing strategies that promote sustainable economic growth, support green innovation, and address social equity issues.

11.2.4. Climate Change and Extreme Events

Climate change and the increasing frequency of extreme weather events are critical factors in future scenarios. The impacts of climate change, such as rising sea levels, more intense heat waves, and altered precipitation patterns, can profoundly affect ecosystems, economies, and communities. Future scenarios must account for these potential impacts and explore strategies for adaptation and resilience. For example, coastal cities may need to invest in infrastructure improvements to protect against flooding, while agricultural systems may require changes to cope with altered growing conditions. Understanding the potential range of climate-related impacts helps develop proactive measures to mitigate risks and enhance resilience (Planton et al., 2008).

11.2.5. Global and Local Dynamics

Foresight for environmental transformation must consider both global and local dynamics. Global trends, such as international climate agreements and global market shifts, influence local environmental policies and practices. Conversely, local actions and innovations can contribute to global environmental goals. Future scenarios should examine how global trends might affect local conditions and vice versa. For example, international efforts to reduce plastic pollution may drive local initiatives to improve waste management and recycling. Similarly, local innovations in sustainable agriculture can inform broader strategies for food security and environmental conservation (Moon et al., 2015). Integrating global and local perspectives helps create more comprehensive and effective strategies for environmental transformation.

11.2.6. Stakeholder Engagement and Collaboration

Effective environmental transformation requires the engagement and collaboration of diverse stakeholders, including governments, businesses, communities, and individuals. Future scenarios should consider the roles and contributions of various stakeholders and explore how they can work together to achieve common goals. Collaborative approaches, such as multi-stakeholder partnerships and participatory decision-making processes, can enhance the effectiveness of environmental policies and actions. Stakeholders can develop more inclusive and innovative solutions to environmental challenges by involving various perspectives and expertise (Desai, 2018).

Foresight and future scenarios are integral to navigating the complexities of environmental transformation. By utilizing scenario planning, embracing emerging technologies, anticipating socio-economic shifts, and considering both global and local dynamics, stakeholders can better prepare for the future and implement effective strategies for sustainability. Engaging diverse stakeholders and incorporating their insights further enhances the ability to address environmental challenges and achieve long-term transformation. As we look ahead, foresight will continue to play a crucial role in shaping our planet's sustainable and resilient future.

11.3. Emerging Technologies for Sustainable Development

Emerging technologies play a transformative role in advancing sustainable development by addressing critical environmental, social, and economic challenges. These innovations offer new solutions and opportunities to enhance resource efficiency, reduce environmental impacts, and promote equitable growth (Ludovico de Almeida et al., 2013).

11.3.1. Renewable Energy Technologies

Renewable energy technologies are at the forefront of efforts to transition to a low-carbon economy. Solar and wind energy have seen significant advancements, making them more efficient and cost-effective. For example, innovations in photovoltaic cells, such as perovskite solar cells, promise higher energy conversion efficiencies at lower costs. Similarly, advances in wind turbine design and offshore wind farms are expanding the potential for harnessing wind energy. These technologies reduce reliance on fossil fuels, lower greenhouse gas emissions, and contribute to energy security. Additionally, energy storage solutions, such as advanced batteries and grid-scale storage systems, are crucial for managing the intermittent nature of renewable energy and ensuring a stable energy supply (Küfeoğlu, 2022).

11.3.2. Artificial Intelligence and Machine Learning

Artificial intelligence (AI) and machine learning are revolutionizing various sectors by enabling more precise data analysis and decision-making. In environmental management, AI algorithms can analyze large datasets to predict environmental changes, optimize resource use, and enhance conservation efforts. For instance, AI is used in wildlife monitoring to track animal populations and detect poaching activities. In agriculture, machine learning models help optimize crop yields by analyzing weather patterns, soil conditions, and pest infestations. AI-driven solutions also support energy management by predicting energy demand and improving the efficiency of renewable energy systems (Khanzode and Sarode, 2020). These technologies enhance our ability to address complex environmental challenges and make informed decisions for sustainable development.

11.3.3. Biotechnology and Environmental Restoration

Biotechnology offers innovative solutions for environmental restoration and sustainability. Genetic engineering and synthetic biology are being used to develop plants and microorganisms that can address environmental issues such as soil degradation, pollution, and climate change. For example, genetically modified plants with enhanced nutrient uptake can improve soil health and fertility. Bioremediation techniques utilize microorganisms to clean up polluted soils and water bodies, breaking down harmful contaminants and restoring ecological balance. Additionally, biotechnology is advancing the development of sustainable biofuels and biodegradable materials, reducing reliance on fossil fuels and minimizing waste (Gavrilescu, 2010).

11.3.4. Smart Cities and Urban Technologies

The concept of smart cities involves integrating technology into urban infrastructure to improve efficiency, sustainability, and quality of life. Smart city technologies include sensors, IoT devices, and data analytics that enable real-time monitoring and management of urban systems. For example, smart grids optimize energy distribution and reduce energy consumption, while intelligent transportation systems improve traffic flow and reduce emissions. Additionally, smart water management systems help monitor and conserve water resources, and green building technologies enhance energy efficiency in construction (Yigitcanlar, 2015). By leveraging these technologies, cities can become more resilient, reduce their environmental footprint, and provide better services to their residents.

11.3.5. Circular Economy Innovations

Emerging technologies are also supporting the shift towards a circular economy, which emphasize reducing waste and maximizing resource use. Innovations in recycling and waste management technologies improve the efficiency of material recovery and recycling processes. For instance, advanced sorting technologies and automated recycling systems enhance the separation and processing of recyclable materials. Additionally, new materials and design approaches, such as biodegradable plastics and modular product designs, contribute to reducing waste and extending product lifecycles (Suchek et al., 2021) These technologies help create closed-loop systems where resources are continuously reused and recycled, minimizing environmental impacts and promoting sustainability.

11.3.6. Blockchain Technology

Blockchain technology offers potential solutions for enhancing transparency, traceability, and accountability in various sectors. In environmental sustainability, blockchain can track and verify the origins of sustainably sourced materials, such as conflict-free minerals or certified organic products. It also enables the creation of decentralized platforms for carbon credit trading, ensuring that emissions reductions are accurately recorded and verified (Yaga et al., 2019). Furthermore, blockchain-based systems can improve supply chain management by providing real-time visibility into the movement and handling of goods, reducing inefficiencies, and promoting sustainable practices.

11.3.7. Water Purification and Management Technologies

Advancements in water purification and management technologies are crucial for addressing water scarcity and improving access to clean water. Innovative technologies, such as advanced filtration systems, desalination methods, and water recycling technologies, are enhancing the efficiency and effectiveness of water treatment. For example, membrane technologies and nanotechnology are being used to develop more efficient water purification systems that can remove contaminants and provide potable water from various sources. Additionally, smart water management systems leverage sensors and data analytics to optimize water usage, detect leaks, and manage water resources more effectively (Coca-Prados and Gutiérrez-Cervelló, 2010).

Emerging technologies drive significant progress towards sustainable development by offering innovative solutions to environmental, social, and economic challenges. From advancements in renewable energy and AI to biotechnology and smart city technologies, these innovations enhance our ability to manage resources efficiently, reduce environmental impacts, and promote equitable growth. By integrating these technologies into strategies for sustainable development, we can address pressing global issues and build a more resilient and sustainable future for all.

11.4. Collaborative Global Efforts for Planetary Health

The concept of planetary health emphasizes the interconnectedness of human well-being and the health of the natural environment. Addressing environmental degradation, climate change, and resource depletion challenges requires collaborative efforts that transcend national borders and involve diverse stakeholders (Wardani et al., 2024).

11.4.1. International Agreements and Frameworks

International agreements and frameworks are central to fostering global cooperation for planetary health. Agreements like the Paris Agreement on climate change represent collective commitments by countries to limit global warming and reduce greenhouse gas emissions. This accord, adopted in 2015, outlines targets for reducing emissions, adapting to climate impacts, and providing financial support to developing nations. Similarly, the Convention on Biological Diversity (CBD) aims to conserve biodiversity, promote sustainable use of biological resources, and ensure equitable sharing of benefits from genetic resources (Matz-Lück, 2012).

These global frameworks create a common platform for countries to align their policies and actions with shared environmental goals. They also facilitate regular reporting, assessment, and negotiation processes to enhance collective efforts and address emerging challenges.

11.4.2. Multilateral Environmental Organizations

Multilateral environmental organizations are crucial in coordinating global efforts for planetary health. Organizations such as the United Nations Environment Programme (UNEP), the World Health Organization (WHO), and the International Union for Conservation of Nature (IUCN) provide guidance, research, and support for environmental and health initiatives. UNEP, for example, leads global efforts to address environmental issues, promote sustainable development, and facilitate international cooperation on environmental challenges (Kim, 2013).

These organizations often work with governments, non-governmental organizations (NGOs), and other stakeholders to implement projects, conduct research, and advocate for policies that support planetary health. They also provide platforms for knowledge sharing and collaboration across sectors and regions.

11.4.3. Cross-Sectoral Partnerships

Addressing planetary health requires collaboration across various sectors, including government, business, academia, and civil society. Cross-sectoral partnerships combine diverse expertise and resources to tackle complex environmental and health challenges. For instance, the United Nations Foundation launched the Global Alliance for Clean Cookstoves, bringing together governments, businesses, and NGOs to promote clean cooking technologies and improve air quality and health in developing countries (Ordonez-Ponce, 2021).

Similarly, partnerships between corporations and environmental organizations can drive sustainable practices and innovation. For example, the Ellen MacArthur Foundation collaborates with businesses to advance the circular economy, promoting resource efficiency and waste reduction (Grace, 2017). These partnerships leverage the strengths of different sectors to achieve shared goals and amplify the impact of their efforts.

11.4.4. Community-Based and Local Initiatives

Local and community-based initiatives are vital components of global efforts for planetary health. Grassroots organizations and local communities often lead innovative projects that address environmental and health issues specific to their regions. For example, community-led reforestation projects, such as those in Kenya and Madagascar, contribute to biodiversity conservation, climate mitigation, and local livelihoods (Lempa et al., 2008).

Supporting and scaling up these initiatives can enhance global efforts by incorporating local knowledge, addressing specific needs, and fostering community engagement. Collaborative networks that connect local initiatives with global platforms can help share best practices, access resources, and amplify the impact of grassroots efforts.

11.4.5. Scientific Collaboration and Research Networks

Scientific collaboration and research networks are critical in advancing knowledge and solutions for planetary health. International research collaborations, such as those coordinated by the Intergovernmental Panel on Climate Change (IPCC) and the International Society for Environmental Epidemiology (ISEE), provide valuable insights into environmental and health interactions and inform policy decisions (Kyvik and Reymert, 2017).

These networks facilitate the sharing of data, methodologies, and findings across borders, enhancing our understanding of complex environmental issues and developing evidence-based solutions. Collaborative research initiatives also foster innovation and capacity building by bringing together scientists, practitioners, and policymakers from different regions and disciplines.

11.4.6. Global Campaigns and Advocacy

Global campaigns and advocacy efforts raise awareness and mobilize action on critical environmental and health issues. Campaigns like the "Fridays for Future" climate strikes, initiated by young activists, and the "Beat Plastic Pollution" campaign by UNEP highlight the urgent need for environmental action and influence policy changes at national and international levels (Hall, 2022).

Advocacy efforts, including the public, policymakers, and businesses, engage a broad audience to drive collective action and policy reform. These campaigns often leverage social media, public demonstrations, and media outreach to amplify their messages and mobilize support for planetary health.

Collaborative global efforts are essential for advancing planetary health and addressing the complex challenges of environmental degradation, climate change, and resource depletion. International agreements, multilateral organizations, cross-sectoral partnerships, community-based initiatives, scientific collaborations, and global campaigns all contribute to a unified approach to sustainability. By fostering cooperation and leveraging diverse expertise and resources, we can work together to protect and restore the health of our planet and ensure a sustainable future for all.

11.5. The Role of Education in Environmental Transformation

Education is a pivotal element in driving environmental transformation. It empowers individuals and communities with the knowledge, skills, and values to address environmental challenges and promote sustainable practices. The role of education in environmental transformation spans various levels and approaches, influencing everything from individual behavior to policy development and systemic change (Alam and Dewi, 2024).

11.5.1. Raising Awareness and Understanding

Education is crucial in raising awareness about environmental issues and fostering a deeper understanding of ecological systems. By integrating environmental education into school curricula and public awareness campaigns, individuals learn about the impacts of human activities on the environment and the importance of conservation. Topics such as climate change, biodiversity loss, and pollution are covered to help people recognize the urgency of these issues and the need for action (Kilkelly et al., 2021). Awareness programs and educational initiatives, such as Earth Day events and environmental documentaries, also build a more informed and engaged public.

11.5.2. Promoting Sustainable Practices and Behavior

Educational programs are instrumental in promoting sustainable practices and behavior at the individual and community levels. Schools and universities offer courses on sustainability, resource management, and environmental ethics, equipping students with practical knowledge and skills. Programs such as waste reduction, energy conservation, and water management are often included in curricula to encourage students to adopt environmentally friendly habits (McKenzie-Mohr, 2000). Additionally, educational campaigns and workshops for adults and businesses help spread knowledge about sustainable practices, such as reducing plastic use, supporting renewable energy, and making sustainable purchasing choices.

11.5.3. Encouraging Innovation and Research

Education fosters innovation and research that contribute to environmental transformation. Academic institutions and research organizations conduct studies and develop technologies aimed at addressing environmental challenges. Educational programs in environmental science, engineering, and technology prepare students to engage in cutting-edge research and develop innovative solutions. For example, universities may focus on research related to renewable energy technologies, waste management systems, or sustainable agriculture practices (Guertler et al., 2020). By supporting academic research and encouraging scientific inquiry, education drives advancements that can lead to more effective and sustainable solutions.

11.5.4. Building Capacity for Environmental Leadership

Education is essential for developing the next generation of environmental leaders and advocates. Leadership programs, training courses, and mentorship opportunities help individuals acquire the skills and knowledge to lead environmental initiatives and influence policy. Environmental leadership academies, community organizing workshops, and advocacy training prepare individuals to engage in activism, policy-making, and community engagement (Boeske, 2023). Education empowers individuals to drive environmental change and contribute to sustainability efforts by nurturing leadership qualities and providing tools for effective advocacy.

11.5.5. Integrating Indigenous and Local Knowledge

Incorporating indigenous and local knowledge into environmental education enriches understanding and practices related to sustainability. Many indigenous communities have traditional ecological knowledge that offers valuable insights into sustainable resource management, biodiversity conservation, and climate adaptation. Educational programs that include this knowledge help bridge the gap between traditional practices and modern science, fostering a more holistic approach to environmental management (Loch and Riechers, 2021). Collaborations between indigenous communities, educators, and researchers can enhance the relevance and effectiveness of environmental education by integrating diverse perspectives and solutions.

11.5.6. Supporting Policy Development and Advocacy

Education supports environmental policy development and advocacy by providing policymakers, advocates, and stakeholders with evidence-based information and training. Educational institutions often conduct research and produce reports that inform policy decisions and contribute to public discourse on environmental issues. Training programs and workshops for policymakers and advocacy groups help build expertise and develop effective policy implementation and lobbying strategies. Educating decision-makers and advocates with knowledge and skills enhances the capacity to address environmental challenges through informed and effective policies (Blagescu and Young, 2006; Jackson, 2014; Stockdale et al., 2006).

11.5.7. Fostering Community Engagement and Action

Education fosters community engagement and action by involving individuals in local environmental initiatives and projects. Community-based education programs, such as citizen science projects, environmental workshops, and local sustainability initiatives, encourage residents to participate in hands-on activities and contribute to environmental stewardship. Educational efforts focusing on community resilience, local conservation projects, and sustainable development empower communities to take collective action and address environmental issues specific to their regions (Russell et al., 2023; Shakesprere et al., 2021; Barnes and Marlatt, 2022).

Education is a cornerstone of environmental transformation, playing a multifaceted role in raising awareness, promoting sustainable practices, encouraging innovation, and building leadership. By integrating ecological education into various levels and sectors, society can equip individuals with the knowledge and skills to address environmental challenges and drive positive change. Through formal education, community programs, or policy support, education empowers people to contribute to a sustainable future and fosters a more informed and proactive approach to environmental stewardship.

Chapter 12

Conclusion

This book on environmentalism transformation has explored a comprehensive range of topics essential for understanding and accelerating the shift toward a sustainable future. Beginning with a historical overview of environmentalism, we traced the development of the movement from its early roots to its current form, providing context for the need for transformation. Defining environmental transformation and highlighting its urgency underscored the need for immediate and effective action to address global environmental challenges.

Examining the theoretical foundations of environmentalism, we established a framework for understanding the principles guiding environmental thought and practice. Identifying the drivers of transformation, including technological innovation, policy shifts, and socio-economic factors, illuminated the forces shaping the future of environmentalism. The role of key stakeholders—governments, businesses, NGOs, and communities—was analyzed to highlight their contributions and responsibilities in advancing environmental goals.

Pathways for environmental transformation were discussed, focusing on practical strategies such as sustainable development, circular economy models, and renewable energy transitions. Environmental knowledge systems and innovation were emphasized as crucial for developing effective solutions and policies. The intersection of environmentalism with social transformation revealed the broader impacts of environmental issues on equity, justice, and community well-being.

Challenges to environmental transformation, including technological limitations, resource scarcity, and global inequality, were addressed. Examining environmentalism in a globalized world showcased the interconnected nature of these issues. Case studies provided real-world examples of successful environmental initiatives, offering valuable insights and lessons.

Looking to the future, the book highlighted emerging theories and practices that can drive continued progress towards a sustainable planet. Drawing from my 50 years of academic and activist experience, this book aims to bridge the knowledge gap and offer new perspectives that can inspire and guide further action. I hope the insights and strategies presented here will contribute to accelerating the transformation needed to address the planetary crisis effectively.

Environmentalism and transformation are intertwined, requiring a multifaceted approach to address global challenges effectively. Education, technological innovation, and collaborative global efforts are essential for fostering sustainable development. To advance environmental transformation, it is crucial to address environmental challenges effectively, it is crucial to strengthen education by integrating environmental awareness and sustainability principles across all levels of learning. Promoting innovation is another key step, which requires supporting research and the development of green technologies and sustainable solutions. Enhancing collaboration by fostering cross-sectoral partnerships and international cooperation is vital for collective action on global environmental issues. Additionally, empowering communities by engaging local and indigenous knowledge ensures that environmental strategies are inclusive, context-specific, and effective. Together, these approaches can create a robust framework for long-term environmental sustainability.

Environmentalism and transformation are critical in addressing the interconnected challenges of climate change, resource depletion, and environmental degradation. Key insights into this relationship highlight the need for comprehensive and integrated approaches to foster sustainability.

12.1. The Role of Education

Education is fundamental in driving environmental transformation. By embedding environmental issues into curricula and public awareness campaigns, education raises awareness, promotes sustainable behaviors, and fosters innovation (Lahijanian, 2011). Educational programs equip individuals with the knowledge and skills necessary for effective environmental stewardship and policy advocacy.

12.2. Technological Innovations

Emerging technologies are pivotal in advancing sustainability. Innovations in renewable energy, such as solar and wind power, are reducing reliance on fossil fuels. AI and machine learning enhance data analysis and resource management, while biotechnology offers solutions for environmental restoration and sustainable agriculture (Anderson, 2022). These technologies contribute to reducing environmental impacts and promoting resource efficiency.

12.3. Collaborative Efforts

Global cooperation is essential for addressing environmental challenges. International agreements like the Paris Agreement and the Convention on Biological Diversity provide frameworks for collective action. Multilateral organizations and cross-sectoral partnerships leverage diverse expertise and resources to tackle complex issues (Foster, 2002). Community-based initiatives and grassroots movements are crucial in local and global environmental efforts.

12.4. Systemic Change

Addressing environmental challenges requires systemic change. This includes transitioning to a circular economy, advancing renewable energy transitions, and integrating environmental considerations into all aspects of policy and practice (Barbosa, 2009). Emphasizing equity and justice ensures that environmental transformation is inclusive and addresses disparities.

12.5. Future Directions

Looking ahead, fostering sustainable development involves strengthening education, supporting technological innovation, enhancing global cooperation, and empowering communities (Mertig and Dunlap, 2001). Focusing on these areas can drive meaningful progress toward a sustainable and resilient future, addressing current and emerging environmental challenges.

These insights underline the importance of a coordinated, multi-faceted approach to environmentalism and transformation, emphasizing the need for collective action and systemic change to achieve long-term sustainability.

Chapter 13

Policy Recommendations for Accelerating Environmental Transformation

Governments should prioritize adopting renewable energy to accelerate environmental transformation by incentivizing solar, wind, and other clean technologies. Implementing stringent regulations to reduce carbon emissions and enhance energy efficiency is essential. Strengthening circular economy frameworks that promote recycling, waste reduction, and sustainable production will minimize environmental impact. Policymakers should also ensure that environmental justice is integrated by addressing the needs of marginalized communities most affected by climate change. Investing in research, innovation, and education, coupled with international cooperation, will create scalable, equitable solutions that drive rapid and long-lasting environmental progress.

13.1. Strengthening Climate Policies and Commitments

Governments should enhance their climate policies by setting ambitious greenhouse gas reduction targets and implementing stringent emission controls. Adopting carbon pricing mechanisms, such as carbon taxes or cap-and-trade systems, can incentivize businesses to reduce their carbon footprint and invest in clean technologies (Pickering et al., 2022). Additionally, integrating climate considerations into all policy-making sectors—transportation, energy, and agriculture—ensures a comprehensive approach to reducing emissions and promoting sustainability (Gomez-Echeverri, 2018). Regularly updating and strengthening national climate action plans, in line with international agreements like the Paris Agreement, will ensure continued progress and accountability.

13.2. Promoting Renewable Energy and Energy Efficiency

Accelerating the transition to renewable energy sources is critical for reducing reliance on fossil fuels and mitigating climate change. Governments should support developing and deploying renewable energy technologies through subsidies, tax incentives, and research funding. Investments in energy storage solutions, smart grids, and energy-efficient infrastructure are also essential for optimizing energy use and integrating renewable sources into the grid (Cantore et al., 2017). Implementing energy efficiency standards for buildings, appliances, and industrial processes can further reduce energy consumption and emissions.

13.3. Supporting Sustainable Agriculture and Land Use

Sustainable agriculture and land use practices are crucial for environmental conservation and food security. Policies should promote practices such as organic farming, agroforestry, and conservation tillage, which enhance soil health, conserve water, and reduce greenhouse gas emissions. Incentivizing land conservation and reforestation projects helps restore ecosystems and mitigate climate change (Andriyani et al., 2017). Additionally, integrating climate resilience into agricultural policies ensures that farming practices can adapt to changing climate conditions and support long-term sustainability.

13.4. Advancing the Circular Economy and Waste Management

Transitioning to a circular economy model, where resources are reused, recycled, and minimized, can significantly reduce waste and environmental impacts. Policymakers should support circular economy initiatives by promoting product design for longevity and recyclability, implementing extended producer responsibility programs, and investing in recycling infrastructure. Encouraging businesses to adopt circular practices and reducing single-use plastics can also contribute to minimizing waste (Ashokkumar et al., 2022). Effective waste management policies that prioritize recycling, composting, and responsible disposal are essential for reducing the environmental footprint of waste.

13.5. Enhancing Environmental Justice and Equity

Ensuring that environmental transformation is equitable and inclusive is vital for addressing systemic inequalities. Policies should prioritize environmental justice by addressing disparities in exposure to environmental hazards and ensuring that marginalized communities benefit from sustainability initiatives. Engaging communities in decision-making processes and incorporating local knowledge can enhance the effectiveness and fairness of environmental policies (Petts, 2005). Supporting community-based projects and initiatives, especially in underserved areas, helps to empower local stakeholders and promote equitable access to environmental benefits.

13.6. Fostering Innovation and Research

Investing in research and innovation is key to developing new solutions for environmental challenges. Governments should increase funding for environmental research, support public-private partnerships, and encourage collaboration between academia, industry, and government. Promoting innovation in clean energy technologies, sustainable materials, and environmental monitoring can drive progress and address emerging issues (Ferrari et al., 2022). Policies that support intellectual property rights and technology transfer can facilitate the dissemination of innovative solutions and accelerate their adoption.

13.7. Strengthening International Cooperation

Global environmental challenges require coordinated international efforts. Policymakers should strengthen international cooperation through agreements and initiatives that address transboundary issues such as climate change, biodiversity loss, and pollution. Supporting international organizations and frameworks, such as the United Nations Framework Convention on Climate Change (UNFCCC) and the Intergovernmental Science-Policy Platform on Biodiversity and Ecosystem Services (IPBES), enhances global collaboration and knowledge sharing (Javed and Chattu, 2020). Additionally, promoting technology and knowledge exchange between countries can support capacity building and accelerate environmental transformation.

Implementing these policy recommendations will accelerate environmental transformation and support the transition to a more sustainable and resilient future. By strengthening climate policies, promoting renewable energy, supporting sustainable practices, advancing circular economy models, ensuring environmental justice, fostering innovation, and enhancing international cooperation, policymakers can address pressing environmental challenges and drive meaningful progress toward sustainability.

13.8. Future Research Directions and Opportunities

The urgency of environmental transformation necessitates a forward-looking approach to research that can accelerate progress and address pressing challenges (da Cruz et al., 2021). Future research should focus on several key areas to drive effective, timely environmental change.

13.8.1. Advanced Climate Modeling and Predictive Analytics

Research in advanced climate modeling and predictive analytics is essential for understanding and anticipating the impacts of climate change. Improved models can provide more accurate projections of climate scenarios, enabling better planning and mitigation strategies. Research should focus on enhancing the precision of climate forecasts, incorporating more variables and feedback mechanisms, and developing robust tools for decision-making (Reddy and Sureshbabu, 2019). Integrating machine learning and big data analytics into climate models can enhance predictive capabilities and inform adaptive strategies.

13.8.2. Innovations in Renewable Energy Technologies

Developing new and improved renewable energy technologies is crucial for accelerating the transition from fossil fuels. Future research should explore innovative solutions in solar, wind, geothermal, and hydropower technologies and energy storage systems such as advanced batteries and hydrogen fuel cells. Research into hybrid systems that combine renewable sources and optimize their efficiency is also important (Engel-Cox, 2020). Additionally, exploring decentralized and community-based energy systems can enhance energy access and resilience.

13.8.3. Sustainable Agriculture and Food Systems

Research into sustainable agriculture and food systems is vital for addressing food security and environmental impacts. Future studies should focus on developing climate-resilient crops, improving soil health through regenerative practices, and reducing the environmental footprint of food production. Research into plant-based and lab-grown foods and innovations in vertical farming and precision agriculture can contribute to more sustainable and efficient food systems (Barrios et al., 2020). Understanding the socio-economic impacts of these innovations on rural communities and smallholder farmers is also crucial.

13.8.4. The Circular Economy and Waste Reduction

Exploring new models for circular economy and waste reduction is essential for minimizing resource use and environmental impacts. Future research should investigate methods for improving recycling technologies, designing products for durability and recyclability, and developing new materials that are biodegradable or easier to process (Gómez-Prado et al., 2022). Research into waste-to-resource technologies, such as converting waste into energy or valuable materials, can also reduce waste and promote circularity.

13.8.5. Environmental Justice and Equity

Addressing environmental justice and equity requires targeted research to understand and mitigate disparities in environmental impacts and access to resources. Future research should examine environmental policies' social, economic, and health effects on marginalized communities and identify strategies to promote equitable outcomes (Warriner et al., 2001). Studies should also focus on integrating indigenous and local knowledge into environmental decision-making and policy frameworks to ensure that diverse perspectives and needs are considered.

13.8.6. Ecosystem Resilience and Restoration

Research into ecosystem resilience and restoration is crucial for mitigating the impacts of environmental degradation and climate change. Future studies should explore methods for restoring degraded ecosystems, enhancing biodiversity, and improving ecosystem services. Research into the role of ecosystems in carbon sequestration, water purification, and climate regulation can provide valuable insights for conservation and restoration efforts (Gillson et al., 2021). Understanding ecosystem interactions and human activities can also inform more effective management practices.

13.8.7. Policy and Governance Innovations

Innovative research in policy and governance can support more effective environmental management and transformation. Future studies should focus on evaluating the effectiveness of existing environmental policies, exploring new governance models, and identifying best practices for integrating environmental considerations into policymaking (Moore and Hartley, 2010). Research into stakeholder engagement and participatory governance can enhance the inclusivity and effectiveness of environmental policies.

13.8.8. Behavioral and Social Change

Understanding behavioral and social change drivers is critical for promoting sustainable practices and achieving environmental goals. Future research should explore the factors influencing individual and collective behaviors related to environmental conservation, consumption, and advocacy. Studies should also investigate effective communication and education strategies to foster greater public engagement and support for environmental initiatives (Glenn, 2004; Tankard and Paluck, 2016).

Accelerating environmental transformation requires a multifaceted research agenda that addresses key challenges and opportunities. Future research can drive meaningful progress toward a sustainable and resilient future by focusing on advanced climate modeling, renewable energy innovations, sustainable agriculture, circular economy practices, environmental justice, ecosystem resilience, policy innovations, and behavioral change. The urgency of these issues underscores the need for continued investment in research and collaboration across disciplines to address the complex and interconnected challenges facing our planet.

About the Author

Professor Medani P. Bhandari, PhD
Akamai University, USA, Sumy State University, Ukraine and Gandaki University, Nepal

Dr. Medani P. Bhandari is a widely recognized humanitarian, professor, author, and editor who has significantly contributed to social and environmental sciences. With expertise in social equity, inequality, inclusion, innovation, feminism, sustainability, climate change, and social and environmental policies and management, Dr. Bhandari has dedicated his career to these important areas. In addition to his extensive scholarly work, Dr. Bhandari is a poet, essayist, and advocate for the environment and social causes. His expertise expands to global and international environmental politics and justice, climate change, sustainable development, public/social policy, the non-profit sector, renewable energy, nature, culture, and power. He has gained field experience across diverse regions, including Asia, Africa, North America, Western Europe, Australia, Japan, and the Middle East. Dr. Bhandari has authored and co-edited several books, including Green Web-II: Standards and Perspectives from the IUCN, 2nd Edition, Getting the Climate Science Facts Right: The Role of the IPCC, Reducing Inequalities Towards Sustainable Development Goals: Multilevel Approach, Educational Transformation, Economic Inequality – Trends, Traps, and Trade-offs, Social Inequality as a Global Challenge, The Unbeatable Challenges of Inequality, Perspectives on Sociological Theories, Methodological Debates, and Organizational Sociology, The Past, Present, and Future Directions of Migration/Immigration, Theories, and Practices of Global Journeys, Transforming Futures – Navigating Society, Environment, and Governance, Live and Let Others Live – In Reference to Sustainability and Environment Conservation, Navigating Sustainability: A Comparative Analysis of Micro vs. Macro Strategies in the least Developed Countries, The Nexus of Climate Change and Land-use – Global Scenario with Reference to Nepal, Women and Society etc.. Social Inequality- Past, Present, and Future (in which 15 African Countries' case studies are included), etc. Furthermore, Dr. Bhandari\'s creative writing encompasses a vast collection of poems and essays. He has published three volumes of poetry in collaboration with Prajita Bhandari. Dr. Bhandari\'s multifaceted contributions to academia, humanitarianism, and creative writing have solidified his reputation as an esteemed figure in social and environmental sciences, profoundly impacting social equity, sustainability, climate change, and beyond. Professor Bhandari holds five master's degrees in social and environmental sciences and a PhD in sociology. For 45 years, he has been actively involved in environmental conservation and humanitarian efforts through nonprofit and academic platforms. Guided by the philosophy that all living and non-living beings are interconnected ("Vasudhaiva Kutumbakam") and the principle of "Live and Let Others Live," his work reflects a deep commitment to the well-being of our planet and its inhabitants.

He serves as an International Program Director at the New York-based NGO Atlantic State Legal Foundation in Syracuse, NY; a professor at Akamai University, USA, and Sumy State University, Ukraine; and an advisor to Gandaki University, Nepal. He is also the UNEP Civil Society Regional Facilitator for North America and the Lead Author of UNEP GEO 7. He is also the editor-in-chief of Strategic Planning for Energy and the Environment, at River Publishers. He is an editor for several publications from Nova Science Publishers, Inc.

https://www.amazon.com/Books-Medani-P-Bhandari/s?rh=n%3A283155%2Cp_27%3AMedani+P.+Bhandari
https://novapublishers.com/?s=medani+bhandari&post_type=product (books link)
https://english.dcnepal.com/category/opinion/ (opinions)
https://www.linkedin.com/in/medani-bhandari-ph-d-8611a018/

Glossary

Access to Resources: The ability of individuals, communities, or nations to obtain necessary resources such as water, food, energy, and land, essential for survival and development.

Addressing Inequality: Involves efforts to reduce disparities in income, opportunities, and social status between individuals and groups, often focusing on marginalized populations.

Africa: The second-largest continent in the world, rich in biodiversity, culture, and natural resources, facing challenges related to development, climate change, and conservation.

Agriculture Technology: The application of scientific knowledge, tools, and innovations to improve farming practices, crop yields, and sustainability, including automation, precision farming, and biotechnology.

Agroforestry: A land management system that combines trees or shrubs with crops and/or livestock farming, enhancing biodiversity, soil health, and ecosystem resilience.

AI and Machine Learning: Artificial Intelligence (AI) refers to systems that mimic human intelligence, while machine learning is a subset of AI that enables computers to learn and improve from experience without explicit programming.

Anthropocentrism: A worldview that places human beings at the center of consideration, often prioritizing human needs and desires over ecological and environmental concerns.

Asia: The largest continent, home to diverse cultures, ecosystems, and economies, facing unique environmental challenges like rapid industrialization, air pollution, and biodiversity loss.

Barriers and Strategies: Obstacles that prevent achieving desired goals, along with the methods and plans developed to overcome these obstacles in areas such as environmental management, policy, and social change.

Basel Convention: An international treaty designed to reduce the movements of hazardous waste between nations, specifically to prevent waste transfer from developed to less-developed countries.

Behavioral and Psychological Barriers: Mental and emotional obstacles that prevent individuals from adopting more sustainable behaviors, such as habits, attitudes, beliefs, and cognitive biases.

Big Data and AI in Monitoring: The use of large data sets and artificial intelligence to monitor environmental conditions, track changes, and predict future trends, contributing to better decision-making and resource management.

Biodiversity Conservation: The protection, preservation, and sustainable management of biological diversity, including species, ecosystems, and genetic diversity, to maintain ecological balance.

Biology and Biodiversity: Biology is the study of living organisms, while biodiversity refers to the variety and variability of life on Earth, crucial for ecosystem resilience and health.

Biotechnology: The use of living organisms, cells, and biological systems to develop technologies and products that improve agriculture, medicine, and industry.

Blockchain: A decentralized digital ledger technology that records transactions across multiple computers, providing transparency and security in fields like supply chains and environmental monitoring.

Building Environmental Leadership: Developing skills, knowledge, and networks to lead efforts in environmental conservation, sustainability, and policy advocacy, fostering strong environmental governance.

Capabilities Approach in Environmentalism: A framework that emphasizes expanding individuals' and communities' capabilities to achieve environmental sustainability and justice, often aligned with the work of Amartya Sen and Martha Nussbaum.

Capacity Building: Enhancing the abilities of individuals, organizations, and societies to perform functions, solve problems, and achieve goals, particularly in sustainable development and environmental management.

Carbon Neutrality: Achieving a balance between emitting carbon dioxide and absorbing it from the atmosphere, often through reducing emissions and investing in carbon offset projects.

Carrying Capacity Concept: The maximum number of individuals or organisms that an environment can support without causing long-term degradation of natural resources and ecosystems.

Challenges and Opportunities: The difficulties and potential advantages faced when addressing issues such as environmental sustainability, climate change, or social justice.

Circular Economy: An economic system aimed at minimizing waste and making the most of resources by keeping products and materials in use for as long as possible through recycling, reusing, and regenerating natural systems.

Civil Society: Non-governmental organizations, community groups, and other grassroots entities that play a role in advocating for social, political, and environmental change.

Climate Action Plans: Strategies and policies developed by governments, organizations, and communities to mitigate and adapt to the effects of climate change.

Climate Change: The long-term shift in global or regional climate patterns, primarily due to human activities such as fossil fuel combustion, deforestation, and industrial activities.

Climate Change Adaptation: Adjustments in systems, practices, and policies to minimize the damage caused by climate change and take advantage of any potential opportunities.

Collaboration and Partnerships: Cooperative efforts between governments, organizations, communities, and individuals to address complex issues such as climate change and environmental conservation.

Community Empowerment: Enabling communities to take control of their own development, through increased access to resources, decision-making power, and participation in governance processes.

Community Engagement: Active participation of local communities in planning, decision-making, and implementation of initiatives that affect them, often in the context of environmental conservation or development projects.

Conservation Agriculture: A set of soil management practices that minimize the disruption of the soil's structure, composition, and biodiversity, aimed at sustainable agriculture and environmental conservation.

Core Concepts: Fundamental ideas or principles that form the foundation of a field of study or practice, such as sustainability, biodiversity, and ecological resilience in environmentalism.

Corporate Responsibility: The obligation of businesses to act ethically and contribute to economic development while improving the quality of life of their employees, communities, and the environment.

Cross-Sectoral Partnerships: Collaborations that span different sectors, such as government, business, and civil society, to address global challenges like environmental sustainability and social inequality.

Cultural Drivers: Social and cultural factors that influence behaviors, attitudes, and practices, particularly in relation to environmental conservation and sustainability.

Decarbonization: The process of reducing carbon dioxide emissions from human activities, particularly in industries like energy, transportation, and agriculture, to mitigate climate change.

Deep Ecology: An environmental philosophy that promotes the inherent worth of all living beings regardless of their utility to humans, advocating for a radical shift in the human relationship with nature.

Deforestation: The large-scale removal of forests, primarily due to logging, agriculture, and development, leading to habitat loss, biodiversity decline, and increased carbon emissions.

Democratization: The process of making systems, institutions, and knowledge more accessible and participatory, often with a focus on equity, inclusivity, and the distribution of power.

Disparities in Resource Availability: Inequalities in access to natural resources, such as water, land, and energy, often linked to socio-economic, political, or geographic factors.

Drivers of Degradation: Factors that contribute to environmental degradation, including overconsumption, deforestation, pollution, and unsustainable land use practices.

Ecofeminism: A movement that links environmentalism and feminism, arguing that the exploitation of nature and the oppression of women are interconnected and must be addressed together.

Ecological Capabilities: The potential of ecosystems to maintain biodiversity, productivity, and resilience in the face of environmental stresses and changes.

Ecological Economics: A transdisciplinary field that addresses the relationships between ecosystems and economic systems, emphasizing sustainability, resource use, and long-term environmental health.

Ecosystem Services: The benefits that humans derive from ecosystems, including clean air and water, food, climate regulation, and pollination, which are essential for survival and well-being.

Education's Role: The importance of education in fostering awareness, knowledge, and skills to promote sustainability, environmental stewardship, and informed decision-making.

Educational Technology: The use of digital tools, resources, and systems to enhance learning and education, including online learning platforms, digital classrooms, and interactive multimedia.

Emerging Technologies: New and innovative technologies that have the potential to significantly impact society, such as artificial intelligence, biotechnology, blockchain, and renewable energy technologies.

Energy Transitions: The process of shifting from fossil fuels to renewable energy sources, such as wind, solar, and hydropower, to achieve sustainable energy systems and reduce carbon emissions.

Environmental Degradation: The deterioration of the natural environment through the depletion of resources, loss of biodiversity, and pollution, often as a result of human activities.

Environmental Ethics: A branch of philosophy that explores the moral relationship between humans and the natural world, focusing on issues such as sustainability, conservation, and the intrinsic value of nature.

Environmental Impact: The effect that human activities or natural events have on the environment, including changes to ecosystems, pollution levels, and biodiversity.

Environmental Justice: The fair treatment and meaningful involvement of all people, regardless of race, color, or income, in environmental policymaking and the distribution of environmental benefits and burdens.

Environmental Knowledge: The understanding of environmental systems, processes, and issues, which includes both scientific knowledge and indigenous or traditional ecological knowledge.

Environmental Management: The process of managing human interactions with the environment to minimize harm and ensure sustainable resource use and ecosystem health.

Environmental Monitoring: The systematic collection of data on environmental conditions, such as air and water quality, biodiversity, and climate, to track changes and inform decision-making.

Environmental Movement: A social and political movement that advocates for the protection and preservation of the environment through activism, policy changes, and public awareness.

Environmental Organizations: Groups, often non-governmental, that focus on environmental protection, conservation, and sustainability, such as Greenpeace, WWF, and local community organizations.

Environmental Problems: Issues that arise from human activities or natural processes that negatively impact the environment, including pollution, habitat destruction, and climate change.

Environmental Restoration: The process of returning ecosystems to their original state or improving degraded environments through activities like reforestation, habitat restoration, and pollution cleanup.

Environmental Science: The interdisciplinary study of the natural environment, including the physical, chemical, and biological processes that shape ecosystems and the human impact on these systems.

Environmental Standards: Regulatory guidelines that set limits on pollutants or activities to protect environmental quality and public health, such as air quality standards and water safety guidelines.

Environmental Stewardship: The responsible management and care of the environment, ensuring that natural resources are sustainably used and preserved for future generations.

Environmental Transformation: Significant changes in the structure and function of environmental systems due to human or natural influences, often focusing on shifting towards sustainability and resilience.

Equity and Access: The fair distribution of resources and opportunities, ensuring that all individuals and communities have equal access to basic needs like water, food, and education.

Equity and Justice Concerns: Issues related to the fair treatment of individuals and communities, particularly those who have been historically marginalized or disadvantaged, in the distribution of resources and decision-making.

Ethical Concerns: Moral considerations or dilemmas that arise in decision-making, particularly regarding the well-being of people, animals, ecosystems, and future generations.

Extreme Events: Severe weather or environmental occurrences, such as hurricanes, floods, droughts, and wildfires, which are often intensified by climate change and can cause significant damage to ecosystems and communities.

Financial Pathways: Strategies and models for financing sustainable development, climate action, and environmental conservation, including green bonds, carbon pricing, and sustainable investment.

Foresight and Future Scenarios: The process of anticipating and preparing for future challenges and opportunities through the development of predictive models and scenario planning.

Fossil Fuel Depletion: The gradual exhaustion of non-renewable energy sources like coal, oil, and natural gas due to overuse and unsustainable extraction.

Four Quadrants of Integral Theory: A framework developed by Ken Wilber that divides human experience into four dimensions (individual interior, individual exterior, collective interior, and collective exterior) to better understand and address complex problems holistically.

Future Directions: Potential pathways and strategies for addressing global challenges such as climate change, social inequality, and technological advancements, focusing on long-term sustainability and resilience.

Future Research Directions: Areas of scientific or academic inquiry that need further investigation to advance understanding and solve pressing issues, particularly in environmental science, technology, and social equity.

Gaia Hypothesis: A scientific theory proposed by James Lovelock that views Earth as a self-regulating, living organism, where biological and geological processes work together to maintain conditions for life.

Gender and Intersectionality: The study of how different aspects of a person's identity, such as gender, race, and class, intersect and contribute to experiences of privilege or discrimination, particularly in the context of environmental justice and equity.

Global Campaigns: International efforts, often led by governments, NGOs, or grassroots movements, to address global issues such as climate change, biodiversity loss, and human rights.

Global Collaboration: Cooperative efforts between countries, organizations, and institutions worldwide to solve common challenges like environmental degradation, climate change, and sustainable development.

Global Cooperation: The process by which nations and international organizations work together to address global challenges, share resources, and develop collective solutions.

Global Environmental Crises: Large-scale environmental challenges that have global impacts, such as climate change, biodiversity loss, deforestation, and pollution, requiring urgent action and cooperation.

Global Environmental Governance: The collective efforts of governments, international organizations, and non-state actors to manage and protect the global environment through policies, agreements, and institutions.

Global Inequality: The unequal distribution of wealth, resources, and opportunities between different countries and regions, often exacerbated by historical, economic, and environmental factors.

Global Trade and Consumption: The global exchange of goods and services and the associated environmental and social impacts, including resource depletion, pollution, and labor exploitation.

Globalization: The process of increasing interconnectedness and interdependence between countries and people through trade, technology, culture, and communication, which has significant economic, social, and environmental implications.

Governance: The structures, policies, and processes by which decisions are made and authority is exercised in a society, including government institutions, laws, and informal networks.

Governments: Institutions at local, national, or international levels that have the authority to create and enforce laws, manage public resources, and shape policies for societal development.

Grassroots Activism: Community-driven movements and actions that advocate for social, political, or environmental change, often starting at the local level and growing to influence broader policy and governance.

Great Green Wall Initiative: An African-led project aimed at combating desertification in the Sahel region by planting a wall of trees across the continent to restore degraded lands and enhance livelihoods.

Green Building Standards: Guidelines and certifications that promote environmentally responsible and resource-efficient construction practices, such as LEED (Leadership in Energy and Environmental Design).

Green Economy: An economic system that aims to reduce environmental risks and ecological scarcities, while improving human well-being and social equity, often through sustainable practices and green technologies.

Green Transitions: The process of shifting from unsustainable practices to environmentally sustainable ones, especially in industries like energy, agriculture, and manufacturing, to reduce environmental impact.

Holistic Environmentalism: An approach to environmentalism that considers the interconnectedness of social, economic, cultural, and ecological factors in addressing environmental challenges.

Human-Induced Drivers: Activities and behaviors by humans that contribute to environmental changes, such as deforestation, industrialization, and pollution, leading to problems like climate change and biodiversity loss.

Inclusivity and Equity: Ensuring that all people, regardless of their background or identity, have equal access to resources, opportunities, and decision-making processes, particularly in the context of environmental justice and sustainability.

Indigenous Knowledge: Traditional knowledge held by indigenous communities, encompassing ecological understanding, cultural practices, and land management techniques passed down through generations, often playing a vital role in conservation and sustainable development.

Indigenous Land Stewardship: The practices and philosophies of indigenous peoples in managing and caring for land and ecosystems, emphasizing sustainable use, biodiversity protection, and a deep connection with nature.

Innovation and Green Technologies: The development and application of new technologies aimed at reducing environmental impact, improving sustainability, and addressing climate change, including renewable energy, green materials, and eco-friendly industrial processes.

Institutional and Governance Pathways: The structures and processes through which institutions and governments develop, implement, and enforce policies, often with a focus on sustainable development and environmental management.

International Agreements: Treaties, accords, or conventions between nations to address global issues like climate change, biodiversity, and pollution, such as the Paris Agreement and the Convention on Biological Diversity.

IPCC (Intergovernmental Panel on Climate Change): A United Nations body tasked with assessing the science related to climate change, providing governments with scientific data to develop informed climate policies and actions.

Knowledge Sharing: The exchange of information, skills, and expertise across individuals, communities, organizations, and governments to promote learning, innovation, and collaboration in solving complex problems.

Lifestyle Changes: Shifts in individual or community behavior and consumption patterns to reduce environmental impact, including adopting sustainable diets, using public transportation, and conserving energy.

Local Actors: Individuals, organizations, or institutions that operate within a specific region or community and play key roles in local decision-making, development, and implementation of sustainable practices.

Local Governments: Municipal or regional government bodies responsible for managing public services, resources, and regulations at the community level, often leading local sustainability efforts.

Local Knowledge: Community-specific understanding and expertise derived from lived experience and observation of the local environment, often used to complement scientific knowledge in environmental management.

Long-Term Risks: Potential threats or challenges that may impact ecosystems, societies, or economies over an extended period, such as climate change, resource depletion, or biodiversity loss.

Loss of Biodiversity: The reduction or extinction of species in ecosystems due to human activities like deforestation, habitat destruction, pollution, and climate change, which diminishes ecosystem services and resilience.

Multilateral and Bilateral Cooperation: Collaboration between two or more countries (bilateral) or multiple nations (multilateral) to address shared environmental, economic, or political issues.

National Governments: Centralized governing bodies responsible for formulating and enforcing laws, policies, and regulations at the national level, often playing a critical role in environmental and sustainability policies.

Natural and Human-induced Drivers: Factors that cause environmental changes, where natural drivers include volcanic activity or climate patterns, while human-induced drivers include deforestation, industrialization, and pollution.

NGOs (Non-Governmental Organizations): Independent organizations that work on various issues such as environmental conservation, human rights, and development, often playing a key role in advocacy, education, and implementation of sustainable practices.

North America: A region encompassing countries like the United States, Canada, and Mexico, playing a significant role in global environmental issues, policy-making, and resource management.

Opportunities for Businesses: The potential for companies to innovate, develop new markets, and contribute to sustainability by adopting green technologies, reducing emissions, and aligning with environmental goals.

Organic Farming: An agricultural practice that avoids synthetic chemicals and genetically modified organisms, focusing on natural processes and biodiversity to enhance soil health and reduce environmental harm.

Overview of the SDGs: A broad summary of the United Nations Sustainable Development Goals (SDGs), which aim to address global challenges like poverty, inequality, climate change, and environmental degradation by 2030.

Policy and Governance: The processes by which laws, regulations, and strategies are developed and implemented by governments to guide societal and environmental outcomes.

Policy Development: The creation and formulation of strategies and regulations by governments, institutions, or organizations to address specific issues such as climate change, biodiversity, or sustainable development.

Policy Frameworks: Structured sets of guidelines and principles that inform the development and implementation of policies, often outlining objectives, strategies, and regulatory mechanisms.

Policy Makers: Individuals or groups in government, business, or organizations responsible for creating, influencing, and implementing policies that address societal issues, including environmental sustainability.

Political Ecology: An interdisciplinary approach that studies the relationships between politics, economics, and environmental issues, focusing on how power dynamics influence environmental outcomes and resource access.

Promoting Sustainable Practices: Encouraging the adoption of behaviors, technologies, and policies that minimize environmental impact, conserve resources, and foster long-term sustainability.

Protected Areas: Designated regions that are legally preserved to conserve biodiversity, ecosystems, and cultural heritage, such as national parks, wildlife reserves, and marine protected areas.

Public Participation: The involvement of citizens in decision-making processes, particularly in environmental policies and actions, ensuring that communities have a voice in shaping sustainable futures.

Raising Awareness: Efforts to inform and educate the public about important environmental issues, fostering understanding and encouraging action to address challenges like climate change and biodiversity loss.

Renewable Energy: Energy generated from natural, replenishable sources such as wind, solar, hydropower, and geothermal, which provide alternatives to fossil fuels and help mitigate climate change.

Resilience Theory: A framework that explores how ecosystems, communities, or systems can absorb disturbances, adapt, and maintain functionality, particularly in the face of environmental or social challenges.

Resource Scarcity: The lack or limited availability of essential natural resources, such as water, energy, and minerals, often leading to competition, conflict, and environmental degradation.

Role of Policy: The influence and impact that laws, regulations, and strategies have on shaping societal behaviors, economic systems, and environmental outcomes, often critical in promoting sustainability.

Scenario Planning and Modeling: The use of simulations and projections to explore possible future outcomes based on different variables, helping organizations and governments prepare for uncertainties and develop adaptive strategies.

Science-Policy Interfaces: The relationships and interactions between scientific research and policy-making, ensuring that decisions are informed by the best available evidence and knowledge.

Scientific Collaboration: Cooperative efforts between scientists across disciplines, institutions, and countries to address complex global challenges, such as climate change, biodiversity loss, and public health.

Social Movements: Collective actions by groups of people advocating for societal change, often focused on issues like environmental justice, human rights, or economic reform.

Social Transformation: Significant changes in the structure, values, and systems of a society, often driven by movements for environmental, social, or political justice, aiming for long-term positive impacts.

Socio-Economic Inequality: The unequal distribution of wealth, opportunities, and resources within a society, often contributing to social injustices and environmental disparities.

Socio-Political Drivers: Social and political factors that influence environmental policies and decisions, such as public opinion, government priorities, political movements, and international relations.

Soil Degradation: The decline in soil health and fertility due to unsustainable agricultural practices, deforestation, or pollution, leading to reduced agricultural productivity and ecosystem damage.

South Asia: A region comprising countries like India, Nepal, Bangladesh, and Sri Lanka, facing significant environmental challenges such as water scarcity, air pollution, and climate vulnerability.

Stakeholder Engagement: The process of involving diverse groups, including governments, businesses, communities, and NGOs, in decision-making and planning to ensure inclusive and effective solutions.

Sustainable Agricultural Practices: Farming methods that promote environmental health, biodiversity, and long-term productivity, such as crop rotation, organic farming, and agroforestry.

Sustainable Development: Development that meets the needs of the present without compromising the ability of future generations to meet their own needs, balancing economic, social, and environmental priorities.

Sustainable Future: A vision of society in which economic growth, social equity, and environmental protection are achieved in a balanced and enduring manner, ensuring long-term well-being for all.

Systems Thinking: An approach to understanding complex problems by examining the relationships and interactions between different parts of a system, recognizing that changes in one area can impact others.

Technological Innovation: The development of new technologies and improvements to existing ones, driving progress in areas like sustainability, resource management, and environmental protection.

The Future of Environmentalism: The evolving strategies and philosophies aimed at addressing environmental challenges, focusing on more inclusive, resilient, and sustainable approaches for the future.

Water Scarcity: The lack of sufficient freshwater resources to meet the needs of populations, ecosystems, and industries, often exacerbated by climate change, overuse, and pollution.

References

Abbas, Q., Yousaf, B., Ali, M. U., Munir, M. A. M., El-Naggar, A., Rinklebe, J., and Naushad, M. (2020). Transformation pathways and fate of engineered nanoparticles (ENPs) in distinct interactive environmental compartments: A review. *Environment International*, *138*, 105646.

Abdmouleh, Z., Alammari, R. A., and Gastli, A. (2015). Review of policies encouraging renewable energy integration and best practices. *Renewable and Sustainable Energy Reviews*, *45*, 249-262.

Abeling, T., Daschkeit, A., Mahrenholz, P., and Schauser, I. (2018). Resilience—A Useful Approach for Climate Adaptation? *Urban disaster resilience and security: Addressing risks in societies*, 461-471.

Adabre, M. A., Chan, A. P., and Darko, A. (2022). Interactive effects of institutional, economic, social and environmental barriers on sustainable housing in a developing country. *Building and Environment*, *207*, 108487.

Adams, R. M. (1989). Global climate change and agriculture: an economic perspective. *American journal of agricultural economics*, *71*(5), 1272-1279.

Adams, W. M. (2005). Green development theory?: Environmentalism and sustainable development. In *Power of development* (pp. 85-96). *Routledge.*

Adamson, D. (2010). Community empowerment: Identifying the barriers to "purposeful" citizen participation. *International Journal of Sociology and Social Policy, 30(3/4), 114-126.*

Adeleke, J. A., Moodley, D., Rens, G., and Adewumi, A. O. (2017). Integrating statistical machine learning in a semantic sensor web for proactive monitoring and control. *Sensors*, *17*(4), 807.

Adelman, S. (2018). The sustainable development goals, anthropocentrism and neoliberalism. In *Sustainable development goals* (pp. 15-40*)*. Edward Elgar Publishing.

Adibfar, A. (2013). The best desalination technology for the Persian Gulf. In *Creating a Sustainable Ecology Using Technology-Driven Solutions* (pp. 326-334*). IGI Global.*

Afshar, N., English, D. R., and Milne, R. L. (2021). Factors explaining socio-economic inequalities in cancer survival: a systematic review. *Cancer Control, 28,* 10732748211011956.

Agyeman, J., Bullard, R. D., and Evans, B. (2002). Exploring the nexus: Bringing together sustainability, environmental justice and equity. *Space and Polity*, *6*(1), 77-90.

Aït-El-Hadjait, S. (2017). *The Ongoing Technological System* (Vol. 11). John Wiley and Sons.

Akhtar, M. K. (2011). A system dynamics based integrated assessment modeling of global-regional climate change: a model for analyzing the behavior of the social-energy-economy-climate system. *The University of Western Ontario (Canada).*

Al Haija, A. A. (2011). Jordan: Tourism and conflict with local communities. *Habitat International*, *35*(1), 93-100.

Alam, A. A., and Dewi, A. E. R. (2024). Integration of Education in Human Resource Development: Transformation of the Human Resource Management Paradigm. *Maneggio, 1(3), 1-9.*

Alberti, M. (1996). Measuring urban sustainability. *Environmental impact assessment review, 16(4-6), 381-424.*

Alexander, T., Burnette, C. B., Cory, H., McHale, S., and Simone, M. (2024). The need for more inclusive measurement to advance equity in eating disorders prevention. *Eating Disorders*, 1-19.

Ali, S. A., Elsaid, S. A., Ateya, A. A., ElAffendi, M., and El-Latif, A. A. A. (2023). Enabling Technologies for Next-Generation Smart Cities: A Comprehensive Review and Research Directions. *Future Internet*, *15*(12), 398.

Allam, Z., and Cheshmehzangi, A. (2024). Technological Innovation and Sustainable Transitions. In *Sustainable Futures and Green New Deals* (pp. 55-79). Cham: Springer Nature Switzerland.

Allen, M. L., Allen, M. M., Cumming, D., and Johan, S. (2021). Comparative capitalisms and energy transitions: Renewable energy in the European Union. *British Journal of Management*, *32*(3), 611-629.

Al-Shetwi, A. Q. (2022). Sustainable development of renewable energy integrated power sector: Trends, environmental impacts, and recent challenges. *Science of The Total Environment, 822, 153645.*

Al-Shorbaji, N., and Alhuwail, D. (2022). Health informatics in the Middle East and North Africa. In *Roadmap to Successful Digital Health Ecosystems* (pp. 375-397). Academic Press.

Altmann, M., Eisenreich, S., Lehner, D., Moser, S., Neidl, T., Rüscher, V., and Vogeler, T. (2013). Global inequality and poverty in perspectives of geography. *Multicultural Education and Technology Journal*, *7*(2/3), 127-150.

Alvarez, C. H., and Evans, C. R. (2021). Intersectional environmental justice and population health inequalities: A novel approach. *Social Science and Medicine*.

Álvarez, L., and Coolsaet, B. (2020). Decolonizing environmental justice studies: a Latin American perspective. *Capitalism Nature socialism*, *31*(2), 50-69.

Amorim-Maia, A. T., Anguelovski, I., Chu, E., and Connolly, J. (2022). Intersectional climate justice: A conceptual pathway for bridging adaptation planning, transformative action, and social equity. *Urban climate*, *41*, 101053.

Anand, P. B. (2014). Sustainability and the Capability Approach: From Theory to Practice?. In *The capability approach: From theory to practice* (pp. 118-147). London: Palgrave Macmillan UK.

Anderson, K. (1996). *Environmental standards and international trade* (p. 30). University of Adelaide, Centre for International Economic Studies.

Anderson, T. (2022). Schoppa, R. Keith. The twentieth century: a world history. *CHOICE: Current Reviews for Academic Libraries*, *59*(10), 1254-1255.

Andriyani, I., Jourdain, D., Lidon, B., Soni, P., and Kartiwa, B. (2017). Upland farming system erosion yields and their constraints to change for sustainable agricultural conservation practices: A case study of land use and land cover (LULC) change in Indonesia. *Land Degradation and Development*, *28*(2), 421-430.

Anesa, P. (2018). Popularization and democratization of knowledge through blawgs. *Iperstoria*, (12).

Arauco, V. P., Gazdar, H., Hevia-Pacheco, P., Kabeer, N., Lenhardt, A., Masood, S. Q., ... and Mariotti, C. (2014). *Strengthening social justice to address intersecting inequalities post-2015*. London, Overseas Development Institute. www.odi.org/sites/odi. org. uk/files/odi-assets/publications-opinion-files/9213. pdf.

Argent, R. M. (2004). An overview of model integration for environmental applications—components, frameworks and semantics. *Environmental Modelling and Software, 19(3), 219-234.*

Argyrou, V. (2005). *The logic of environmentalism: anthropology, ecology and postcoloniality* (Vol. 1). Berghahn Books.

Arkell, B. T. (2018). *Integral ecology as applied to environmental policy, politics and democracy* (Doctoral dissertation, University of Tasmania).

Arkema, K. K., Abramson, S. C., and Dewsbury, B. M. (2006). Marine ecosystem-based management: from characterization to implementation. *Frontiers in Ecology and the Environment*, *4*(10), 525-532.

Aronczyk, M. (2018). Public relations, issue management, and the transformation of American environmentalism, 1948–1992. *Enterprise and Society*, *19*(4), 836-863.

Arora-Jonsson, S. (2014, November). Forty years of gender research and environmental policy: Where do we stand?. In *Women's Studies International Forum* (Vol. 47, pp. 295-308). Pergamon.

Arora-Jonsson, S. (2017). Gender and environmental policy. *Routledge handbook of gender and environment*, 289-303.

Arruda, E. H., Melatto, R. A. P. B., Levy, W., and de Melo Conti, D. (2021). Circular economy: A brief literature review (2015–2020). *Sustainable Operations and Computers*, *2*, 79-86.

Arts, B., and Leroy, P. (2006). Institutional processes in environmental governance: lots of dynamics, not much change?. *Environment and Policy*, *47*, 267.

Asaduzzaman, M., Kaivo-oja, J., Stenvall, J., and Jusi, S. (2016). Strengthening local governance in developing countries: Partnership as an alternative approach. *Public Organization Review*, *16*, 335-356.

Ashokkumar, V., Flora, G., Venkatkarthick, R., SenthilKannan, K., Kuppam, C., Stephy, G. M., ... and Ngamcharussrivichai, C. (2022). Advanced technologies on the sustainable approaches for conversion of organic waste to valuable bioproducts: Emerging circular bioeconomy perspective. *Fuel*, *324*, 124313.

Ategeka, C. (2021). *The Unintended Consequences of Technology: Solutions, Breakthroughs, and the Restart We Need*. John Wiley and Sons.

Awasthi, A., Chauhan, S. S., and Omrani, H. (2011). Application of fuzzy TOPSIS in evaluating sustainable transportation systems. *Expert Systems with Applications*, *38*(10), 12270-12280.

Ayars, P. J. (2009). *The art of leading transformational change*. AuthorHouse.

Aydın, Z. (2016). The state, civil society, and environmentalism. In *Environmentalism in Turkey* (pp. 67-84). Routledge.

Babones, S. J., and Turner, J. H. (2004). Global inequality. *Handbook of Social Problems*, 101-20.

Bäckstrand, K. (2003). Civic science for sustainability: reframing the role of experts, policy-makers and citizens in environmental governance. *Global environmental politics*, *3*(4), 24-41.

Baek, J., Cho, Y., and Koo, W. W. (2009). The environmental consequences of globalization: A country-specific time-series analysis. *Ecological economics*, *68*(8-9), 2255-2264.

Baggett, R. K. (2017). Ethical and Privacy Implications of Technology. *Homeland Security Technologies for the 21st Century*, 17.

Baihakki, M. A., and Qutayan, S. M. S. B. (2023). Ethical issues of artificial intelligence (AI) in the healthcare. *Journal of Science, Technology and Innovation Policy*, *9*(1), 32-38.

Baird, J. (2001). The land ethic. *A Companion to Environmental Philosophy*, 204-17.

Balogh, J. M., and Jámbor, A. (2020). The environmental impacts of agricultural trade: A systematic literature review. *Sustainability*, *12*(3), 1152.

Banerjee, S. B. (2001). Managerial perceptions of corporate environmentalism: Interpretations from industry and strategic implications for organizations. *Journal of management studies*, *38*(4), 489-513.

Banerjee, S. B. (2017). Corporate environmentalism and the greening of strategic marketing: Implications for marketing theory and practice. In *Greener Marketing* (pp. 16-40). *Routledge*.

Bang, M. (2016). A case study: Learning gardens in an urban indigenous community: Expanding the scope of learning. In *Sowing seeds in the city: Human dimensions* (pp. 257-268). Dordrecht: Springer Netherlands.

Barbosa, L. C. (2009). Change by Necessity: Ecological Limits to Capitalism, Climate Change, and Obstacles to Transition to an Environmentally Sustainable Economy. *Climate Change, and Obstacles to Transition to an Environmentally Sustainable Economy* (November 15, 2009).

Barca, S. (2012). On working-class environmentalism: a historical and transnational overview. *Interface: a journal for and about social movements*, *4*(2), 61-80.

Bardoscia, M., Battiston, S., Caccioli, F., and Caldarelli, G. (2017). Pathways towards instability in financial networks. *Nature communications*, *8*(1), 14416.

Barendse, J., Roux, D., Currie, B., Wilson, N., and Fabricius, C. (2016). A broader view of stewardship to achieve conservation and sustainability goals in South Africa. *South African Journal of Science*, *112*(5-6), 1-15.

Barna, C., Zbuchea, A., and Stănescu, S. M. (2023). Social economy enterprises contributing to the circular economy and the green transition in Romania. *CIRIEC-España, revista de economía pública, social y cooperativa*, (107), 47-69.

Barnes, M. E., and Marlatt, R. (2022). From involvement to solidarity: Community engagement to foster culturally-proactive and constructivist pedagogy. *Journal of Curriculum and Pedagogy*, *19*(1), 4-27.

Barnes, M., Newman, J., and Sullivan, H. C. (2007). *Power, participation and political renewal: Case studies in public participation*. Policy Press.

Barrett, S. (2001). Strategic environmental policy and international trade. In *The Economics of International Trade and the Environment* (pp. 83-93). CRC Press.

Barrios, E., Gemmill-Herren, B., Bicksler, A., Siliprandi, E., Brathwaite, R., Moller, S., ... and Tittonell, P. (2020). The 10 Elements of Agroecology: enabling transitions towards sustainable agriculture and food systems through visual narratives. *Ecosystems and People*, *16*(1), 230-247.

Barry, J. (1999). Marxism and ecology. *Marxism and social science*, 259-279.

Barton, P. S., Westgate, M. J., Foster, C. N., Cuddington, K., Hastings, A., O'Loughlin, L. S., ... and Lindenmayer, D. B. (2020). Using ecological niche theory to avoid uninformative biodiversity surrogates. *Ecological Indicators*, *108*, 105692.

Basile, G., and Caputo, F. (2017). Theories and challenges for systems thinking in practice. *Journal of Organisational Transformation and Social Change*, *14*(1), 1-3.

Bassett, E., and Shandas, V. (2010). Innovation and climate action planning: Perspectives from municipal plans. *Journal of the American Planning Association*, *76*(4), 435-450.

Bastiaens, I., and Postnikov, E. (2017). Greening up: the effects of environmental standards in EU and US trade agreements. *Environmental Politics, 26(5), 847-869.*

Batterbury, S., Forsyth, T., and Thomson, K. (1997). Environmental transformations in developing countries: hybrid research and democratic policy. *Geographical journal*, 126-132.

Bauer, F., Hansen, T., and Nilsson, L. J. (2022). Assessing the feasibility of archetypal transition pathways towards carbon neutrality–A comparative analysis of European industries. *Resources, Conservation and Recycling*, *177*, 106015.

Baumgardner, J., and Richards, A. (2005). *Grassroots: A field guide for feminist activism.* Macmillan.

Behnam, S., Cagliano, R., and Grijalvo, M. (2018). How should firms reconcile their open innovation capabilities for incorporating external actors in innovations aimed at sustainable development?. *Journal of Cleaner Production*, *170*, 950-965.

Behnassi, M., Shahid, S. A., and D'silva, J. (2011). *Sustainable agricultural development* (p. 275). Springer, Heidelberg, Berlin.

Bell, K. (2014). Achieving environmental justice. In *Achieving Environmental Justice* (pp. 213-236). *Policy Press.*

Bell, K. (2021). Diversity and inclusion in environmentalism. In *Diversity and Inclusion in Environmentalism* (pp. 1-14). Routledge.

Bennett, E. M., Peterson, G. D., and Gordon, L. J. (2009). Understanding relationships among multiple ecosystem services. *Ecology letters*, *12*(12), 1394-1404.

Bennett, N. J., Blythe, J., Cisneros-Montemayor, A. M., Singh, G. G., and Sumaila, U. R. (2019). Just transformations to sustainability. *Sustainability*, *11*(14), 3881.

Benson, D., Jordan, A., and Smith, L. (2013). Is environmental management really more collaborative? A comparative analysis of putative 'paradigm shifts' in Europe, Australia, and the United States. *Environment and Planning A*, *45*(7), 1695-1712.

Benzidia, S., Makaoui, N., and Bentahar, O. (2021). The impact of big data analytics and artificial intelligence on green supply chain process integration and hospital environmental performance. *Technological forecasting and social change, 165*, 120557.

Berkes, F., Berkes, M. K., and Fast, H. (2007). Collaborative, integrated management in Canada's north: The role of local and traditional knowledge and community-based monitoring. *Coastal management, 35*(1), 143-162.

Bernstein, S. (2001). *The compromise of liberal environmentalism.* Columbia University Press.

Bernstein, S. (2002). Liberal environmentalism and global environmental governance. *Global environmental politics*, 2(3), 1-16.

Bernstein, S. (2004). Legitimacy in global environmental governance. *J. Int'l L and Int'l Rel., 1*, 139.

Bernstein, S. (2004). Legitimacy in global environmental governance. *J. Int'l L and Int'l Rel., 1, 139.*

Best, R. (2017). Switching towards coal or renewable energy? The effects of financial capital on energy transitions. *Energy Economics, 63*, 75-83.

Betts-Davies, S., Barrett, J., Brockway, P., and Norman, J. (2024). Is all inequality reduction equal? Understanding motivations and mechanisms for socio-economic inequality reduction in economic narratives of climate change mitigation. *Energy Research and Social Science, 107*, 103349.

Beveridge, R., and Kern, K. (2013). The Energiewende in Germany: background, developments and future challenges. *Renewable Energy L. and Pol'y Rev., 4*, 3.

Bhandari, M. (2014). Civil society and non-governmental organizations (NGOs) movements in Nepal in terms of social transformation. *The Pacific Journal of Science and Technology, 15*(1), 177-189.

Bhandari, M. (2014a). Is tourism always beneficial? A case study from Masai Mara national reserve, Narok, Kenya. *Pacific Journal of Science and Technology, 15*(1), 458-483.

Bhandari, M. P. (2012). Exploring the International Union for the Conservation of Nature (IUCN's) National Program Development in Biodiversity Conservation: A Comparative Study of India, Pakistan, Nepal, and Bangladesh (Doctoral dissertation, Syracuse University).

Bhandari, M. P. (2018). The role of international organizations in addressing climate change issues and the creation of the Intergovernmental Panel on Climate Change (IPCC). *Adv Agr Environ Sci, 1(1), 19.*

Bhandari, M. P. (2019). Live and let other live-the harmony with nature/living beings-in reference to sustainable development (SD)-is contemporary world's economic and social phenomena is favorable for the sustainability of the planet in reference to India, Nepal, Bangladesh, and Pakistan. *Adv Agr Environ Sci. (2019), 2*(1), 37-57.

Bhandari, M. P. (2019a). BashudaivaKutumbakkam. The entire world is our home and all living beings are our relatives. Why we need to worry about climate change, with reference to pollution problems in the major cities of India, Nepal, Bangladesh and Pakistan. *Adv Agr Environ Sci. (2019), 2*(1), 8-35.

Bhandari, M. P. (2020). Editorial Strategic Planning for Energy and the Environment: Theories, Practice and the Future. *Strategic Planning for Energy and the Environment*, 381-402.

Bhandari, M. P. (2022). Social Inequality as a Global Challenge: Scenario, Impacts, and Consequences. *In Social Inequality as a Global Challenge (pp. 1-25).* River Publishers.

Bhandari, M. P. (2022). The Pathways to Address the Challenges of Inequality. In *Inequality–the unbeatable challenge* (pp. 307-320). River Publishers.

Bhandari, M. P. (2022). *Getting the climate science facts right: the role of the IPCC*. River Publishers.

Bhandari, M. P. (2022). *Green Web-II: Standards and Perspectives from the IUCN Program/Policy Development in Environment Conservation Domain-with reference to India,* Pakistan, Nepal, and Bangladesh. River Publishers.

Bhandari, M. P. (2023). The fundamental principles of social sciences. *Business Ethics and Leadership*, *7*(2), 73-86.

Bhandari, M. P. (2023). *Perspectives on Sociological Theories, Methodological Debates and Organizational Sociology*. River Publishers.

Bhandari, M. P. (2023b). *Perspectives on Sociological Theories, Methodological Debates and Organizational Sociology*. River Publishers.

Bhandari, M. P. (2023c). Past and present of social inequality: analyzing structure and future trends. *Futurity of Social Sciences*, *1*(3), 47-60.

Bhandari, M. P. (2024). Feminisms in Social Sciences. In *Women and Society. IntechOpen.*

Bhandari, M. P. (2024a). *The Nexus of Climate Change and Land-use–Global Scenario with Reference to Nepal*. CRC Press.

Bhandari, M. P. (Ed.). (2024a). *Women and Society*. BoD–Books on Demand.

Bhandari, M. P., and Hanna, S. (Eds.). (2022). *Economic Inequality–Trends, Traps and Trade-offs*. CRC Press.

Bhandari, M. P., and Hanna, S. (Eds.). (2022a). *Inequality–the unbeatable challenge*. CRC Press.

Bhandari, M. P., and Oli, K. P. (2018). The changing roles and impacts of civil society/ NGOs in Nepal. In *Civil Society in the Global South* (pp. 70-87). Routledge.

Bhardwaj, R., Bindra, S., and Srivastava, S. (2021). Sustainability through the lens of dynamic capability: a histographic analysis. *Academy of Marketing Studies Journal*, *25*, 1-6.

Bhattacharya, A. K. (2005). *Eco-tourism and livelihoods: Capacity building for local authorities*. Concept Publishing Company.

Bhattacharyya, A., and Rahman, M. L. (2020). Values, gender and attitudes towards environmental policy: A study of future managers. *Business Strategy and the Environment*, *29*(6), 2514-2527.

Bhattarai, U., Maraseni, T., and Apan, A. (2022). Assay of renewable energy transition: A systematic literature review. *Science of The Total Environment*, *833*, 155159.

Bibri, S. E., Alexandre, A., Sharifi, A., and Krogstie, J. (2023). Environmentally sustainable smart cities and their converging AI, IoT, and big data technologies and solutions: an integrated approach to an extensive literature review. *Energy Informatics*, *6*(1), 9.

Biermann, F., and Pattberg, P. (2008). Global environmental governance: Taking stock, moving forward. *Annual Review of Environment and Resources*, *33*(1), 277-294.

Biermann, F., and Pattberg, P. (Eds.). (2012). *Global environmental governance reconsidered.* MIT Press.

Biesbroek, G. R., Klostermann, J. E., Termeer, C. J., and Kabat, P. (2013). On the nature of barriers to climate change adaptation. *Regional Environmental Change, 13*, 1119-1129.

Biggs, Duan, Alison J. Peel, Christos Astaras, Alexander Braczkowski, Hubert Cheung, Chi-Yeung Choi, Robinson Diotoh Orume et al. (2023). "Governance principles for the wildlife trade to reduce spillover and pandemic risk." *CABI One Health* (2023): ohcs202300013.

Binetti, M. S., Massarelli, C., and Uricchio, V. F. (2024). Machine Learning in Geosciences: A Review of Complex Environmental Monitoring Applications. *Machine Learning and Knowledge Extraction, 6*(2), 1263-1280.

Biondi, V., Iraldo, F., and Meredith, S. (2002). Achieving sustainability through environmental innovation: the role of SMEs. *International Journal of Technology Management, 24*(5-6), 612-626.

Blagescu, M., and Young, J. (2006). *Capacity development for policy advocacy: current thinking and approaches among agencies supporting civil society organizations.* London: Overseas Development Institute.

Blair, R. (2004). Public participation and community development: The role of strategic planning. *Public Administration Quarterly, 28(1), 102-147.*

Blanc, N. (2019). From ordinary environmentalism to the public environment: theoretical reflections based on French and European empirical research. *Ecology and Society, 24*(3).

Blanco-Canqui, H., Lal, R., Blanco-Canqui, H., and Lal, R. (2008). Soil and water conservation. *Principles of soil conservation and management*, 1-19.

Bodansky, D. (1993). The United Nations framework convention on climate change: a commentary. *Yale J. Int'l l., 18*, 451.

Boeske, J. (2023). Leadership towards sustainability: a review of sustainable, sustainability, and environmental leadership. *Sustainability, 15*(16), 12626.

Boillat, S., and Berkes, F. (2013). Perception and interpretation of climate change among Quechua farmers of Bolivia: indigenous knowledge as a resource for adaptive capacity. *Ecology and society, 18*(4).

Boiral, O. (2002). Tacit knowledge and environmental management. *Long range planning, 35*(3), 291-317.

Bollyky, T. J., Cockburn, I. M., and Berndt, E. (2010). Bridging the gap: improving clinical development and the regulatory pathways for health products for neglected diseases. *Clinical trials, 7*(6), 719-734.

Bonaglia, D., Lema, R., de Medina, M. M., and Rabellotti, R. (2024). Digital and Green Transitions: Opportunities and Challenges for Europe and China. *Reverse Dependency: Making Europe's Digital Technological Strengths Indispensable to China*, 184-207.

Boston, P. Q., Mitchell, M. M., Collum, K., and Gravlee, C. C. (2015). Community engagement and health equity. *Practicing Anthropology, 37*(4), 28-32.

Boström, M., and Lidskog, R. (2024). *Environmental Sociology and Social Transformation: Key Issues.* Taylor and Francis.

Bothello, J., and Salles-Djelic, M. L. (2018). Evolving conceptualizations of organizational environmentalism: A path generation account. *Organization Studies*, *39*(1), 93-119.

Bothello, J., and Salles-Djelic, M. L. (2018). Evolving conceptualizations of organizational environmentalism: A path generation account. *Organization Studies*, *39*(1), 93-119.

Bousset, J. P., Skuras, D., Těšitel, J., Marsat, J. B., Petrou, A., Fiallo-Pantziou, E., ... and Bartoš, M. (2007). A decision support system for integrated tourism development: Rethinking tourism policies and management strategies. *Tourism Geographies*, *9*(4), 387-404.

Boyle, J. (2007). Cultural environmentalism and beyond. *Law and Contemporary Problems*, *70*(2), 5-21.

Brand, P., and Thomas, M. (2013). Urban environmentalism: Global change and the mediation of local conflict. *Routledge.*

Brand, U., and Wissen, M. (2017). Social-ecological transformation. *International encyclopedia of geography*, 1-9.

Breitmeier, H. (1997). International organizations and the creation of environmental regimes. *Global governance: Drawing insights from the environmental experience*, 87-114.

Breitmeier, H. (1997). International organizations and the creation of environmental regimes. *Global governance: Drawing insights from the environmental experience*, 87-114.

Brondízio, E. S., Aumeeruddy-Thomas, Y., Bates, P., Carino, J., Fernández-Llamazares, Á., Ferrari, M. F., ... and Shrestha, U. B. (2021). Locally based, regionally manifested, and globally relevant: Indigenous and local knowledge, values, and practices for nature. *Annual Review of Environment and Resources*, *46*(1), 481-509.

Brown Travis, C., and Compton, J. D. (2001). Feminism and health in the decade of behavior. *Psychology of Women Quarterly*, *25*(4), 312-323.

Brown, L. R., Salim, E., Kashiwagi, T., and Takeuchi, K. (1996). The Conceptual Transformation of the Environment toward Global Cooperation for Environmental Protection. *Global Environmental Security: From Protection to Prevention*, 1-69.

Brown, R. D. (2007). The history of wildlife conservation and research in the United States–and implications for the future. *Proc. Taiwan.* Wildl. Assoc, 1-17.

Bruckmeier, K. (2016). Social-ecological transformation. *London: Palgrave Macmillan.*

Brugnach, M., Craps, M., and Dewulf, A. R. P. J. (2017). Including indigenous peoples in climate change mitigation: addressing issues of scale, knowledge and power. *Climatic change*, *140*, 19-32.

Brusseau, M. L., Ramirez-Andreotta, M., Pepper, I. L., and Maximillian, J. (2019). Environmental impacts on human health and well-being. In *Environmental and pollution science* (pp. 477-499). Academic Press.

Buchan, D. (2012). The Energiewende–Germany's gamble. *Oxford Institute for Energy Studies.*

Buchanan, M., and Wozniak-Brown, J. (2023). Barriers to Equity Within Environmental Justice and Climate Justice Grant Programs. *Journal of Climate Resilience and Climate Justice*, *1*, 119-133.

Buchanan, R. (2019). Systems thinking and design thinking: The search for principles in the world we are making. *She Ji: The Journal of Design, Economics, and Innovation*, *5*(2), 85-104.

Buchy, M., and Hoverman, S. (2000). Understanding public participation in forest planning: a review. *Forest Policy and Economics*, *1*(1), 15-25.

Buckingham, S., and Le Masson, V. (Eds.). (2017). *Understanding climate change through gender relations.* Taylor and Francis.

Buechler, S. M. (2016). *Understanding social movements: Theories from the classical era to the present*. Routledge.

Bullard, R. D. (Ed.). (2007). *Growing smarter: Achieving livable communities, environmental justice, and regional equity*. Mit Press.

Bullard, R. D., and Johnson, G. S. (2000). Environmentalism and public policy: Environmental justice: Grassroots activism and its impact on public policy decision making. *Journal of Social Issues*, *56*(3), 555-578.

Butler, C. D., Corvalan, C. F., and Koren, H. S. (2005). Human health, well-being, and global ecological scenarios. *Ecosystems*, *8*, 153-162.

Cabrera, D., and Cabrera, L. (2023). What is systems thinking?. In *Learning, design, and technology: An international compendium of theory, research, practice, and policy* (pp. 1495-1522). Cham: Springer International Publishing.

Cabrera, D., and Cabrera, L. (2023). What is systems thinking?. In *Learning, design, and technology: An international compendium of theory, research, practice, and policy* (pp. 1495-1522). Cham: Springer International Publishing.

Cairns, M. (Ed.). (2007). Voices from the forest: integrating indigenous knowledge into sustainable upland farming. *Earthscan.*

Callicott, J. B., and McRae, J. (Eds.). (2014). *Environmental philosophy in Asian traditions of thought*. SUNY Press.

Cannon, C., Goldsmith, K., and Roux, C. (2019). A self-regulatory model of resource scarcity. *Journal of Consumer Psychology*, *29*(1), 104-127.

Cantore, N., Nussbaumer, P., Wei, M., and Kammen, D. M. (2017). Promoting renewable energy and energy efficiency in Africa: a framework to evaluate employment generation and cost-effectiveness. *Environmental Research Letters*, *12*(3), 035008.

Capano, G., and Jarvis, D. S. (Eds.). (2020). *Convergence and diversity in the governance of higher education: Comparative perspectives*. Cambridge University Press.

Caprotti, F., and Bailey, I. (2014). Making sense of the green economy. *Geografiska annaler: series B, Human geography*, *96*(3), 195-200.

Cardenete, M. A., and Sancho, F. (2012). The role of supply constraints in multiplier analysis. *Economic Systems Research*, *24*(1), 21-34.

Cardinale, B. J., Duffy, J. E., Gonzalez, A., Hooper, D. U., Perrings, C., Venail, P., ... and Naeem, S. (2012). Biodiversity loss and its impact on humanity. *Nature*, *486*(7401), 59-67.

Carlson, J. L., Haffenden, R. A., Bassett, G. W., Buehring, W. A., Collins III, M. J., Folga, S. M., ... and Whitfield, R. G. (2012). *Resilience: Theory and Application* (No. ANL/DIS-12-1). Argonne National Lab. (ANL), Argonne, IL (United States).

Carmichael, J. T., Jenkins, J. C., and Brulle, R. J. (2012). Building environmentalism: The founding of environmental movement organizations in the United States, 1900–2000. *The Sociological Quarterly*, *53*(3), 422-453.

Carnwell, R., and Carson, A. (2005). Understanding partnerships and collaboration. *Effective practice in health and social care*, 4-20.

Carpenter, S. R., Bennett, E. M., and Peterson, G. D. (2006). Scenarios for ecosystem services: an overview. *Ecology and Society*, *11*(1).

Carr, D. S., and Halvorsen, K. (2001). An evaluation of three democratic, community-based approaches to citizen participation: Surveys, conversations with community groups, and community dinners. *Society and Natural Resources*, *14*(2), 107-126.

Cassegård, C. (2017). Eco-Marxism and the critical theory of nature: Two perspectives on ecology and dialectics. *Distinktion: Journal of social theory*, *18*(3), 314-332.

Castaneda, I., Aguilar, C., and Rand, A. (2012). *Measurement and reporting: important elements of gender mainstreaming in environmental policies.* Cornell JL and Pub. Pol'y, 22, 667.

Castree, N. (2006). The future of environmentalism. *Soundings*, *34*, 11-22.

Cavoukian, A., Dix, A., and El Emam, K. (2014). The unintended consequences of privacy paternalism. *Information and Privacy Commissioner of Ontario,* Canada.

Cecot, C., and Hahn, R. W. (2024). Incorporating equity and justice concerns in regulation. *Regulation and Governance*, *18*(1), 99-120.

Ćetković, S., and Buzogány, A. (2016). Varieties of capitalism and clean energy transitions in the European Union: When renewable energy hits different economic logics. *Climate Policy*, *16*(5), 642-657.

Chandra, A., and Idrisova, A. (2011). Convention on Biological Diversity: a review of national challenges and opportunities for implementation. *Biodiversity and Conservation*, *20*, 3295-3316.

Chang, S. J., and Wu, B. (2014). Institutional barriers and industry dynamics. *Strategic Management Journal*, *35*(8), 1103-1123.

Change, C. (1990). The IPCC scientific assessment. *Contribution of Working Group I to the First Assessment Report of the Intergovernmental Panel on Climate Change*, *365*.

ChanLin, L. J. (2008). Technology integration applied to project-based learning in science. *Innovations in education and teaching international, 45(1),* 55-65.

Chaudhary, A., and Brooks, T. M. (2019). National consumption and global trade impacts on biodiversity. *World Development*, *121*, 178-187.

Chaudhry, F. N., and Malik, M. F. (2017). Factors affecting water pollution: a review. *J. Ecosyst. Ecography*, *7*(1), 225-231.

Chen, A., He, H., Wang, J., Li, M., Guan, Q., and Hao, J. (2019). A study on the arable land demand for food security in China. *Sustainability*, *11*(17), 4769.

Chen, J., Beam, A., Saria, S., and Mendonça, E. A. (2019). Potential trade-offs and unintended consequences of AI. *Artificial Intelligence in Health Care: The Hope, the Hype, the Promise, the Peril*, 89.

Chen, W., Zou, W., Zhong, K., and Aliyeva, A. (2023). Machine learning assessment under the development of green technology innovation: A perspective of energy transition. *Renewable Energy*, *214*, 65-73.

Chertow, M. (1997). *Thinking ecologically: The next generation of environmental policy.* Yale University Press.

Chofreh, A. G., and Goni, F. A. (2017). Review of frameworks for sustainability implementation. *Sustainable Development, 25*(3), 180-188.

Choudhury, A. S. B. (2019). Eco-tourism: The mantra for sustainable rural livelihood. *IJRAR-International Journal of Research and Analytical Reviews (IJRAR), 6*(1), 227-229.

Chung, J. Y. (1997). Capitalist development, space and environment in postwar Korea: a regulationist approach (Doctoral dissertation, University of Liverpool).

Clapp, J., and Dauvergne, P. (2011). *Paths to a green world: The political economy of the global environment.* MIT press.

Clark, S. S., and Miles, M. L. (2021). Assessing the integration of environmental justice and sustainability in practice: a literature review. *Sustainability, 13*(20), 11238.

Coates, J. (2003). Exploring the roots of the environmental crisis: Opportunity for social transformation. *Critical Social Work, 4*(1), 1-15.

Coates, P. (2013). *Nature: Western attitudes since ancient times.* John Wiley and Sons.

Coca-Prados, J., and Gutiérrez-Cervelló, G. (Eds.). (2010). *Water purification and management.* Springer.

Coglianese, C. (2001). Social movements, law, and society: The institutionalization of the environmental movement. *University of Pennsylvania Law Review, 150*(1), 85-118.

Cohen, S. (2014). Understanding environmental policy. *Columbia University Press.*

Coleman, S., and Thomas, B. (2017). *Organizational change explained: Case studies on transformational change in organizations.* Kogan Page Publishers.

Cooke, P. (2013). Transition regions: green innovation and economic development. In *Cooperation, Clusters, and Knowledge Transfer: Universities and Firms Towards Regional Competitiveness* (pp. 105-125*).* Berlin, Heidelberg: Springer Berlin Heidelberg.

Cooper, A., Turney, C. S., Palmer, J., Hogg, A., McGlone, M., Wilmshurst, J., ... and Zech, R. (2021). A global environmental crisis 42,000 years ago. *Science, 371*(6531), 811-818.

Cooper, C. M. (2006). 'This is Our Way In': The Civil Society of Environmental NGOs in South-West China1. *Government and Opposition, 41*(1), 109-136.

Corsatea, T. D. (2014). Technological capabilities for innovation activities across Europe: evidence from wind, solar and bioenergy technologies. *Renewable and Sustainable Energy Reviews, 37*, 469-479.

Corvalan, C., Hales, S., and McMichael, A. J. (2005). Ecosystems and human well-being: health synthesis. *World Health Organization.*

Corvellec, H., Stowell, A. F., and Johansson, N. (2022). Critiques of the circular economy. *Journal of Industrial Ecology, 26*(2), 421-432.

Craft, J., and Howlett, M. (2012). Policy formulation, governance shifts and policy influence: Location and content in policy advisory systems. *Journal of Public Policy, 32*(2), 79-98.

Curran, G. (2006). *21st century dissent: Anarchism, anti-globalization and environmentalism.* Springer.

Curtin, C. G., and Parker, J. P. (2014). Foundations of resilience thinking. *Conservation biology*, *28*(4), 912-923.

da Cruz, T. T., Balestieri, J. A. P., de Toledo Silva, J. M., Vilanova, M. R., Oliveira, O. J., and Ávila, I. (2021). Life cycle assessment of carbon capture and storage/utilization: From current state to future research directions and opportunities. *International Journal of Greenhouse Gas Control*, *108*, 103309.

Daily, G. C., and Matson, P. A. (2008). Ecosystem services: from theory to implementation. *Proceedings of the national academy of sciences*, *105*(28), 9455-9456.

Dalziel, P., Saunders, C., and Saunders, J. (2018). *Wellbeing economics: The capabilities approach to prosperity* (p. 196). Springer Nature.

D'amato, D., and Korhonen, J. (2021). Integrating the green economy, circular economy and bioeconomy in a strategic sustainability framework. *Ecological Economics*, *188*, 107143.

Dandabathula, G., Bhardwaj, P., Burra, M., Rao, P. V. P., and Rao, S. S. (2019). Impact assessment of India's Swachh Bharat Mission–Clean India Campaign on acute diarrheal disease outbreaks: Yes, there is a positive change. *Journal of family medicine and primary care*, *8*(3), 1202-1208.

Darko, I. N. (2014). *Environmental stewardship and Indigenous African philosophies: Implications for schooling, and health education in Africa: A case of Ghana*. University of Toronto (Canada).

Datta, P. S. (2019). *Water harvesting for groundwater management: issues, perspectives, scope, and challenges*. John Wiley and Sons.

Davey, I. (2009). Environmentalism of the poor and sustainable development: an appraisal. *JOAAG*, *4*(1), 1-10.

Davey, I. (2009). Environmentalism of the poor and sustainable development: an appraisal. *JOAAG*, *4*(1), 1-10.

Davidson, D. J. (2019). Exnovating for a renewable energy transition. *Nature Energy*, *4*(4), 254-256.

Davoudi, S., Brooks, E., and Mehmood, A. (2013). Evolutionary resilience and strategies for climate adaptation. *Planning Practice and Research*, *28*(3), 307-322.

Dawson, A. (2024). Environmentalism from Below: How Global People's Movements are Leading the Fight for Our Planet. *Haymarket Books.*

Dawson, T. P., Jackson, S. T., House, J. I., Prentice, I. C., and Mace, G. M. (2011). Beyond predictions: biodiversity conservation in a changing climate. *Science*, *332*(6025), 53-58.

Day, R. (2017). A capabilities approach to environmental justice. In *The Routledge handbook of environmental justice* (pp. 124-135). Routledge.

De Jonge, E. (2017). *Spinoza and deep ecology: Challenging traditional approaches to environmentalism.* Routledge.

De Oliveira, J. P., Balaban, O., Doll, C. N., Moreno-Peñaranda, R., Gasparatos, A., Iossifova, D., and Suwa, A. (2011). Cities and biodiversity: Perspectives and governance challenges for implementing the convention on biological diversity (CBD) at the city level. *Biological Conservation*, *144*(5), 1302-1313.

De Ploey, J., Imeson, A., and Oldeman, L. R. (1991). Soil erosion, soil degradation and climatic change. *Land use changes in europe: Processes of change, environmental transformations and future patterns*, 275-292.

de Royston, M. M., Barron, B., Bell, P., Pea, R., Stevens, R., and Goldman, S. (2020). Learning pathways: How learning is culturally organized. In *Handbook of the Cultural Foundations of Learning* (pp. 195-211). *Routledge.*

De Silva, M., Howells, J., Khan, Z., and Meyer, M. (2022). Innovation ambidexterity and public innovation Intermediaries: The mediating role of capabilities. *Journal of Business Research*, *149*, 14-29.

Deakin, M. (2009). A community-based approach to sustainable urban regeneration. *Journal of Urban Technology*, *16*(1), 91-112.

Del Granado, P. C., Resch, G., Holz, F., Welisch, M., Geipel, J., Hartner, M., ... and Ramos, A. (2020). Energy Transition Pathways to a Low-carbon Europe in 2050. *Economics of Energy and Environmental Policy*, *9*(1), 121-136.

del Mar Delgado-Serrano, M., Mistry, J., Matzdorf, B., and Leclerc, G. (2017). Community-based management of environmental challenges in Latin America and the Caribbean. *Ecology and Society*, *22*(1).

Desai, V. M. (2018). Collaborative stakeholder engagement: An integration between theories of organizational legitimacy and learning. *Academy of Management Journal*, *61*(1), 220-244.

DeSombre, E. R. (2010). *Global environmental institutions*. Routledge.

Deubelli, T. M., and Mechler, R. (2021). Perspectives on transformational change in climate risk management and adaptation. *Environmental Research Letters*, *16*(5), 053002.

Dev, I., Ram, A., Kumar, N., Uthappa, A. R., and Arunachalam, A. (2022). Conservation Agriculture in Agroforestry Systems. In *Conservation Agriculture in India* (pp. 285-302). *Routledge.*

Di Chiro, G. (2020). Mobilizing 'intersectionality'in environmental justice research and action in a time of crisis. *Environmental justice*, 316-333.

Dilts, R., Hallbom, T., and Smith, S. (2012). *Beliefs: Pathways to health and well-being*. Crown House Publishing.

Dinar, S. (2011a). Resource scarcity and environmental degradation: analyzing international conflict and cooperation. *Beyond Resource Wars: Scarcity, Environmental Degradation, and International Cooperation*, 3-22.

Dinar, S. (Ed.). (2011). *Beyond resource wars: scarcity, environmental degradation, and international cooperation*. MIT Press.

Dolezal, A., Majano, A. M., Ochs, A., and Palencia, R. (2013). *The way forward for renewable energy in Central America*. WorldWatch Institute, Washington DC. See: www. worldwatch. org/way-forward-renewable-energy-central-america.

Dredge, D. (2017). Institutional and policy support for tourism social entrepreneurship. *Social entrepreneurship and tourism: Philosophy and practice*, 35-55.

Driessen, P. P., Dieperink, C., Van Laerhoven, F., Runhaar, H. A., and Vermeulen, W. J. (2012). Towards a conceptual framework for studying shifts in modes of environmental governance–experiences from the Netherlands. *Environmental policy and governance*, *22*(3), 143-160.

Droz, L., Fricke, M. F., Heroor, N., Jannel, R., Komatsubara, O., Lagasca-Hiloma, C. M. A., ... and Sihombing, H. H. (2024). Environmental philosophy in Asia: Between eco-orientalism and ecological nationalisms. *Environmental Values*, 09632719241276999.

Dryzek, J. S., Hunold, C., Schlosberg, D., Downes, D., and Hernes, H. K. (2002). Environmental transformation of the state: the USA, Norway, Germany and the UK. *Political Studies*, *50*(4), 659-682.

Dudgeon, R. C., and Berkes, F. (2003). Local understandings of the land: Traditional ecological knowledge and indigenous knowledge. In *Nature across cultures: Views of nature and the environment in non-Western cultures* (pp. 75-96). Dordrecht: Springer Netherlands.

Duffy, F. M. (2009). Paradigms, mental models, and mindsets: Triple barriers to transformational change in school systems. *International Journal of Educational Leadership Preparation*, *4*(3), n3.

Dumanski, J., Peiretti, R., Benites, J. R., McGarry, D., and Pieri, C. (2006). The paradigm of conservation agriculture. *Proceedings of world association of soil and water conservation*, *1*(2006), 58-64.

Dwivedi, A. K. (2017). Researches in water pollution: A review. *International Research Journal of Natural and Applied Sciences*, *4*(1), 118-142.

Eddy, P. L. (2010). *Partnerships and collaboration in higher education*: AEHE. John Wiley and Sons.

Edwards, G. A., Reid, L., and Hunter, C. (2016). Environmental justice, capabilities, and the theorization of well-being. *Progress in Human Geography*, *40*(6), 754-769.

Ehler, L. E. (2006). Integrated pest management (IPM): definition, historical development and implementation, and the other IPM. *Pest management science*, *62*(9), 787-789.

Eichsteller, M., Njagi, T., and Nyukuri, E. (2022). The role of agriculture in poverty escapes in Kenya–developing a capabilities approach in the context of climate change. *World Development*, *149*, 105705.

Eisner, M. A. (2007). *Governing the environment: The transformation of environmental regulation*. Boulder, CO: Lynne Rienner Publishers.

Eklund, L. (1999). *From citizen participation towards community empowerment*. Tampere University Press.

El Ansari, W., Phillips, C. J., and Hammick, M. (2001). Collaboration and partnerships: developing the evidence base. *Health and Social Care in the Community*, *9*(4), 215-227.

Elder, M., and Olsen, S. H. (2019). The Design of Environmental Priorities in the SDG s. *Global Policy*, *10*, 70-82.

Elena, B., and Gleb, A. (2013). Optimization, niche and neutral mechanisms in the formation of biodiversity. *American Journal of Life Sciences*, *1*(4), 174-183.

Elisha, O. D. (2021). Resources scarcity and sustainable economic growth. *Int. J. Sci. Res. Educ*, *14*, 43-44.

El-Kholei, A. O., and Al Jayyousi, O. R. (2019). Technology-driven innovative municipalities: The case of Dubai Municipality. In *Technology-driven innovation in Gulf Cooperation Council (GCC) countries: Emerging research and opportunities* (pp. 24-48). IGI Global.

El-Kogali, S. E. T., and Krafft, C. (Eds.). (2019). Expectations and aspirations: A new framework for education in the Middle East and North Africa. *World Bank Publications.*

Elliot, R. (Ed.). (1995). *Environmental ethics.* Oxford University Press.

Elliot, S. (2011). Transdisciplinary perspectives on environmental sustainability: a resource base and framework for IT-enabled business transformation. *Mis quarterly*, 197-236.

Elliott, N. C., Farrell, J. A., Gutierrez, A. P., van Lenteren, J. C., Walton, M. P., and Wratten, S. (1995). *Integrated pest management.* Springer Science and Business Media.

Emadodin, I., and Bork, H. R. (2012). Degradation of soils as a result of long-term human-induced transformation of the environment in Iran: an overview. *Journal of Land Use Science*, *7*(2), 203-219.

Engel-Cox, J. (2020). Innovations in Renewable Energy Technologies, Systems, and Energy Analysis (No. NREL/PR-6A50-76085). *National Renewable Energy Lab.* (NREL), Golden, CO (United States).

Engle, E. W., Larkins, M., Bratman, E., and Higgins, A. K. (2024). Centering diversity, equity, inclusion, and justice in environmental studies and sciences by practicing compassionate pedagogies. *Journal of Environmental Studies and Sciences*, 1-15.

Epstein, B. (1995). Grassroots environmentalism and strategies for social change. *New Political Science*, *16*(1), 1-24.

Esbjörn-Hargens, S. (2005). Integral ecology: The what, who, and how of environmental phenomena. *World futures*, *61*(1-2), 5-49.

Esbjörn-Hargens, S., and Zimmerman, M. E. (2009). *Integral ecology: Uniting multiple perspectives on the natural world.* Shambhala Publications.

Esbjörn-Hargens, S., and Zimmerman, M. E. (2009). *Integral ecology: Uniting multiple perspectives on the natural world.* Shambhala Publications.

Essl, F., Dullinger, S., Genovesi, P., Hulme, P. E., Jeschke, J. M., Katsanevakis, S., ... and Bacher, S. (2019). A conceptual framework for range-expanding species that track human-induced environmental change. *BioScience*, *69*(11), 908-919.

Ester, P., Simões, S., and Vinken, H. (2004). Cultural change and environmentalism: a cross-national approach of mass publics and decision makers. *Ambiente and sociedade*, *7*, 45-66.

Esty, D. C. (2008). Climate change and global environmental governance. *Global Governance*, *14*, 111.

Faivre, N., Fritz, M., Freitas, T., De Boissezon, B., and Vandewoestijne, S. (2017). Nature-Based Solutions in the EU: Innovating with nature to address social, economic and environmental challenges. *Environmental Research*, *159*, 509-518.

Falkner, R. (2021). *Environmentalism and global international society* (Vol. 156). Cambridge University Press.

Fan, C., Xu, J., Natarajan, B. Y., and Mostafavi, A. (2023). Interpretable machine learning learns complex interactions of urban features to understand socio-economic inequality. *Computer-Aided Civil and Infrastructure Engineering*, *38*(14), 2013-2029.

Fankhauser, S. (2017). Adaptation to climate change. *Annual Review of Resource Economics*, *9*(1), 209-230.

Farazmand, A. (Ed.). (2023). *Global encyclopedia of public administration, public policy, and governance.* Springer Nature.

Fath, B. D. (2014). Sustainable systems promote wholeness-extending transformations: The contributions of systems thinking. *Ecological modeling*, *293*, 42-48.

Fealy, G. M., Carney, M., Drennan, J., Treacy, M., Burke, J., O'CONNELL, D. Y. M. P. N. A., ... and Sheerin, F. (2009). Models of initial training and pathways to registration: a selective review of policy in professional regulation. *Journal of Nursing Management*, *17*(6), 730-738.

Febiyanti, W., and Nadlifatin, R. (2024). Transformation of Consumer Behavior Through Smart City Technology: A Literature Review. *Journal of Information Systems and Informatics*, *6*(3), 1517-1534.

Feola, G. (2015). Societal transformation in response to global environmental change: A review of emerging concepts. *Ambio*, *44*(5), 376-390.

Fernandes, J. P., and Guiomar, N. (2016). Environmental ethics: driving factors beneath behavior, discourse and decision-making. *Journal of Agricultural and Environmental Ethics*, *29*, 507-540.

Fernández-Llamazares, Á., Lepofsky, D., Lertzman, K., Armstrong, C. G., Brondizio, E. S., Gavin, M. C., ... and Vaughan, M. B. (2021). Scientists' warning to humanity on threats to indigenous and local knowledge systems. *Journal of Ethnobiology*, *41*(2), 144-169.

Ferrara, P. (2019). Sustainable international relations. Pope Francis' encyclical Laudato Si'and the planetary implications of "integral ecology," *Religions, 10(8), 466.*

Ferrari De Carvalho Teixeira, M. C., Albina Da Silva Pereira, N., Mazzuchetti, R. N., and Lermen, F. H. (2022). Entrepreneurial view: fostering entrepreneurship and innovation in universities by a web-based course. *Computer applications in engineering education*, *30*(5), 1338-1349.

Figge, L., Oebels, K., and Offermans, A. (2017). The effects of globalization on Ecological Footprints: an empirical analysis. *Environment, Development and Sustainability*, *19*, 863-876.

Figueroa, R. M. (2022). Environmental justice. In *The Routledge Companion to environmental ethics* (pp. 767-782). *Routledge.*

Filho, W. L., Shiel, C., and Paço, A. D. (2015). Integrative approaches to environmental sustainability at universities: an overview of challenges and priorities. *Journal of Integrative Environmental Sciences*, *12*(1), 1-14.

Finger, M. (2013). NGOs and transformation: beyond social movement theory. In *Environmental NGOs in world politics* (pp. 48-66). Routledge.

Fischer, F. (1993). Citizen participation and the democratization of policy expertise: From theoretical inquiry to practical cases. *Policy sciences*, *26*(3), 165-187.

Fischer, M. M. (2001). Innovation, knowledge creation and systems of innovation. *The annals of regional science*, *35*, 199-216.

Fischer-Kowalski, M., and Haberl, H. (Eds.). (2007). *Socioecological transitions and global change: Trajectories of social metabolism and land use.* Edward Elgar Publishing.

Fisher, R. M. (2019). *Fear on the run: Towards an Integral Environmentalism.* University of Calgary Digital Repository.

Fitzgerald, J. (2010). *Emerald cities: Urban sustainability and economic development.* Oxford University Press.

Flaskerud, J. H. (2022). COVID-19, health disparities and resource availability. *Issues in Mental Health Nursing, 43*(10), 971-974.

Fletcher, A. J. (2018). More than women and men: A framework for gender and intersectionality research on environmental crisis and conflict. *Water security across the gender divide*, 35-58.

Fonteyne, S., Silva Avendaño, C., Ramos Sanchez, A., Torres Zambrano, J. P., García Dávila, F., Pérez Martínez, Z., ... and Verhulst, N. (2022). Innovating traditional production systems through on-farm conservation agriculture and agroforestry research. *Frontiers in Agronomy, 3*, 787507.

Fors, H., Molin, J. F., Murphy, M. A., and van den Bosch, C. K. (2015). User participation in urban green spaces–For the people or the parks?. *Urban Forestry and Urban Greening, 14*(3), 722-734.

Forsyth, T. (2004). Social movements and environmental democratization in Thailand. *Earthly politics: Local and global in environmental governance*, 195-215.

Forsyth, T. (2012). Democratizing environmental expertise about forests and climate. In *Environmental Governance* (pp. 170-185). Routledge.

Foster, S. (2002). Environmental justice in an era of devolved collaboration. *Harv. Envtl. L. Rev., 26*, 459.

Foweraker, J. (2001). Grassroots movements and political activism in Latin America: A critical comparison of Chile and Brazil. *Journal of Latin American Studies, 33(4), 839-865.*

Fraga, H., Malheiro, A. C., Moutinho-Pereira, J., and Santos, J. A. (2012). An overview of climate change impacts on European viticulture. *Food and Energy Security, 1*(2), 94-110.

Franks, S. J., and Hoffmann, A. A. (2012). Genetics of climate change adaptation. *Annual review of genetics, 46*(1), 185-208.

Free, A., and Barton, N. H. (2007). Do evolution and ecology need the Gaia hypothesis?. *Trends in ecology and evolution, 22*(11), 611-619.

Fu, Y., and Ma, W. (2020). Sustainable urban community development: A case study from the perspective of self-governance and public participation. *Sustainability, 12*(2), 617.

Gaard, G. (1993). Ecofeminism. *In The Routledge Companion to Gender and Animals* (pp. 49-70). Routledge.

Gaard, G. (2017). Critical ecofeminism. *Lexington Books.*

Gabler, C. B., Richey Jr, R. G., and Rapp, A. (2015). Developing an eco-capability through environmental orientation and organizational innovativeness. *Industrial Marketing Management, 45*, 151-161.

Galanis, M. (2023). Corporate law coasting in neutral: from egalitarianism to sustainability, to extinction? In *A Research Agenda for Corporate Law* (pp. 35-55). *Edward Elgar Publishing.*

Gardi, C., Panagos, P., Van Liedekerke, M., Bosco, C., and De Brogniez, D. (2015). Land take and food security: assessment of land take on the agricultural production in Europe. *Journal of Environmental Planning and Management, 58*(5), 898-912.

Gasper, D. (2017). What is the capability approach? Its core, rationale, partners and dangers. In *Development ethics* (pp. 217-241). *Routledge.*

Gatti, R. C. (2011, January). Evolution is a cooperative process: the biodiversity-related niches differentiation theory (BNDT) can explain why. In *Theoretical biology forum* (Vol. 104, No. 1, pp. 35-43).

Gavrilescu, M. (2010). Environmental biotechnology: achievements, opportunities and challenges. *Dynamic biochemistry, process biotechnology and molecular biology, 4*(1), 1-36.

Geels, F. W. (2011). Environmental Innovation and Societal Transitions. *Environmental Innovation and Societal Transitions, 1*, 24-40.

Geels, F. W., Sovacool, B. K., Schwanen, T., and Sorrell, S. (2017). Sociotechnical transitions for deep decarbonization. *Science, 357*(6357), 1242-1244.

Gennari, J. H., Musen, M. A., Fergerson, R. W., Grosso, W. E., Crubézy, M., Eriksson, H., ... and Tu, S. W. (2003). The evolution of Protégé: an environment for knowledge-based systems development. *International Journal of Human-computer studies, 58*(1), 89-123.

Georgeson, L., Maslin, M., and Poessinouw, M. (2017). The global green economy: a review of concepts, definitions, measurement methodologies and their interactions. *Geo: Geography and Environment, 4*(1), e00036.

Georgieva, K. (2019). Environmental Politics and Policy in Bulgaria: Challenges and Constraints to Democratization. In *Beyond Government* (pp. 265-285). *Routledge.*

Ghaith, K. (2024). AI Integration in Cultural Heritage Conservation–Ethical Considerations and the Human Imperative. *International Journal of Emerging and Disruptive Innovation in Education: Visionarium, 2*(1), 6.

Gillson, L., Dirk, C., and Gell, P. (2021). Using long-term data to inform a decision pathway for restoration of ecosystem resilience. *Anthropocene, 36*, 100315.

Gilmurray, J. (2018). Ecology and environmentalism in contemporary sound art (Doctoral dissertation, University of the Arts London).

Giri, A. (1992). Understanding contemporary social movements. *Dialectical anthropology, 17*(1), 35-49.

Gitterman, A. (2009). Ecological perspective. *Social Workers' Desk Reference*, 231.

Glaser, M. A. R. I. O. N., and Glaeser, B. E. R. N. H. A. R. D. (2023). Cross-scale and multi-level analysis of coastal and marine social-ecological systems dynamics. *Coastal Management Revisited: Navigating towards Sustainable Human-Nature Relations*, 241.

Glaser, M., and Glaeser, B. (2014). Towards a framework for cross-scale and multi-level analysis of coastal and marine social-ecological systems dynamics. *Regional Environmental Change, 14*(6), 2039-2052.

Glenn, S. S. (2004). Individual behavior, culture, and social change. *The Behavior Analyst, 27*, 133-151.

Glenn-Lewin, D. C., Peet, R. K., and Veblen, T. T. (Eds.). (1992). *Plant succession: theory and prediction* (Vol. 11). Springer Science and Business Media.

Gliedt, T., Hoicka, C. E., and Jackson, N. (2018). Innovation intermediaries accelerating environmental sustainability transitions. *Journal of cleaner production, 174*, 1247-1261.

Goddard, G., and Farrelly, M. A. (2018). Just transition management: Balancing just outcomes with just processes in Australian renewable energy transitions. *Applied Energy*, *225*, 110-123.

Goel, P. K. (2006). Water pollution: causes, effects and control. *New age international.*

Goffner, D., Sinare, H., and Gordon, L. J. (2019). The Great Green Wall for the Sahara and the Sahel Initiative as an opportunity to enhance resilience in Sahelian landscapes and livelihoods. *Regional Environmental Change*, *19*, 1417-1428.

Gomez-Echeverri, L. (2018). Climate and development: enhancing impact through stronger linkages in the implementation of the Paris Agreement and the Sustainable Development Goals (SDGs). *Philosophical Transactions of the Royal Society A: Mathematical, Physical and Engineering Sciences*, *376*(2119), 20160444.

Gómez-Prado, R., Alvarez-Risco, A., Sánchez-Palomino, J., de las Mercedes Anderson-Seminario, M., and Del-Aguila-Arcentales, S. (2022). Circular economy for waste reduction and carbon footprint. In *Circular Economy: Impact on Carbon and Water Footprint* (pp. 139-159). *Singapore: Springer Singapore.*

González-García, S., Lozano, R. G., Moreira, M. T., Gabarrell, X., i Pons, J. R., Feijoo, G., and Murphy, R. J. (2012). Eco-innovation of a wooden childhood furniture set: an example of environmental solutions in the wood sector. *Science of the Total Environment*, *426*, 318-326.

Gopalakrishnan, B. N., Duggal, T., and Tewary, T. (Eds.). (2021). *Examining the Intersection of Circular Economy, Forestry, and International Trade.* IGI Global.

Görg, C., Brand, U., Haberl, H., Hummel, D., Jahn, T., and Liehr, S. (2017). Challenges for social-ecological transformations: Contributions from social and political ecology. *Sustainability*, *9*(7), 1045.

Grace, R. (2017). Closing the circle: Reshaping how products are conceived and made: Ideo and Ellen MacArthur Foundation create an outline for a New Plastics Economy and launch a Circular Design Guide to help. *Plastics Engineering*, *73*(3), 8-12.

Gracia, J. N., and Koh, H. K. (2011). Promoting environmental justice. *American journal of public health*, *101*(S1), S14-S16.

Granjou, C. (2016). *Environmental changes: The futures of nature*. Elsevier.

Green, G. C., and Barborak, J. (1987). Conservation for Development Success Stories from Central America. *The Commonwealth Forestry Review*, 91-102.

Greenfield, P. M., Keller, H., Fuligni, A., and Maynard, A. (2003). Cultural pathways through universal development. *Annual review of psychology*, *54*(1), 461-490.

Grendstad, G., Selle, P., Stromsnes, K., and Bortne, O. (2006). *Unique environmentalism: A comparative perspective*. Springer Science and Business Media.

Grin, J., Rotmans, J., and Schot, J. (2010). *Transitions to sustainable development: new directions in the study of long term transformative change*. Routledge.

Grizzle, R. E. (1994). Environmentalism should include human ecological needs. *Bioscience*, *44*(4), 263-268.

Gross, J., and Wilson, N. (2020). Cultural democracy: an ecological and capabilities approach. *International journal of cultural policy*, *26*(3), 328-343.

Grumbach, K., Vargas, R. A., Fleisher, P., Aragón, T. J., Chung, L., Chawla, C., ... and Schmidt, L. A. (2017). Achieving health equity through community engagement in translating evidence to policy: the San Francisco Health Improvement Partnership, 2010-2016. *Preventing chronic disease*, *14*(3), E27.

Guertler, M. R., Kriz, A., and Sick, N. (2020). Encouraging and enabling action research in innovation management. *R&D Management*, *50*(3), 380-395.

Guo, L. (2021). Authoritarian decision making at the interface of the state, science and the public: politics of biodiversity conservation and biosafety regulations in China (Doctoral dissertation, University of British Columbia).

Gutés, M. C. (1996). The concept of weak sustainability. *Ecological economics, 17(3), 147-156.*

Haas, P. M. (2017). Coupling science to governance: Straddling the science-policy interface 1. In *The politics of expertise in international organizations* (pp. 54-73). *Routledge.*

Hagen, P. E., and Walls, M. P. (2005). The Stockholm Convention on persistent organic pollutants. *Natural Resources and Environment*, *19*(4), 49-52.

Haila, Y., and Kouki, J. (1994, January). The phenomenon of biodiversity in conservation biology. In *Annales Zoologici Fennici* (pp. 5-18). Finnish Zoological Publishing Board, formed by the Finnish Academy of Sciences, Societas Biologica Fennica Vanamo, Societas pro Fauna et Flora Fennica, and Societas Scientiarum Fennica.

Halecki, W., Stachura, T., Fudała, W., Stec, A., and Kuboń, S. (2023). Assessment and planning of green spaces in urban parks: A review. *Sustainable Cities and Society*, *88*, 104280.

Hall, C. M. (1999). Rethinking collaboration and partnership: A public policy perspective. *Journal of sustainable tourism*, *7*(3-4), 274-289.

Hall, J., Matos, S. V., and Martin, M. J. (2014). Innovation pathways at the base of the pyramid: Establishing technological legitimacy through social attributes. *Technovation*, *34*(5-6), 284-294.

Hall, N. (2022). *Transnational Advocacy in the Digital Era: Think Global, Act Local.* Oxford University Press.

Haluza-DeLay, R. (2008). A theory of practice for social movements: Environmentalism and ecological habitus. *Mobilization: An International Quarterly*, *13*(2), 205-218.

Haluza-DeLay, R. (2008). A theory of practice for social movements: Environmentalism and ecological habitus. *Mobilization: An International Quarterly*, *13*(2), 205-218.

Hamilton, C., Bonneuil, C., and Gemenne, F. (2015). *The Anthropocene and the global environmental crisis: Rethinking modernity in a new epoch* (p. 200). Taylor and Francis.

Handelman, H., and Brynen, R. (2019). Challenges of the developing world. Rowman and Littlefield.

Handl, G. (2012). Declaration of the United Nations conference on the human environment (Stockholm Declaration), 1972 and the Rio Declaration on Environment and Development, 1992. *United Nations Audiovisual Library of International Law*, *11*(6), 1-11.

Hansen, Z. K., and Libecap, G. D. (2004). Small farms, externalities, and the Dust Bowl of the 1930s. *Journal of Political Economy*, *112*(3), 665-694.

Hao-Nan, Z. H. A. N. G., Rong-Rong, S. H. E. N., Xing-Ping, Z. H. A. N. G., Jun-Jie, K. A. N. G., and Jia-Hai, Y. U. A. N. (2022). Implications and pathways of China's carbon neutrality: a review. *Advances in Climate Change Research*, *18*(2), 240.

Haq, G., and Paul, A. (2013). *Environmentalism since 1945*. Routledge.

Harper, S. (2014). Economic and social implications of aging societies. *Science*, *346*(6209), 587-591.

Harris, P. G. (1996). Considerations of equity and international environmental institutions. *Environmental Politics*, *5*(2), 274-301.

Harrison, P. A., Jäger, J., Frantzeskaki, N., and Berry, P. (2019). Understanding high-end climate change: from impacts to co-creating integrated and transformative solutions. *Regional environmental change*, *19*, 621-627.

He, J., Li, Z., Zhang, X., Wang, H., Dong, W., Du, E., ... and Zhang, D. (2022). Towards carbon neutrality: a study on China's long-term low-carbon transition pathways and strategies. *Environmental Science and Ecotechnology*, *9*, 100134.

Healy, S. (2009). Toward an epistemology of public participation. *Journal of Environmental Management*, *90*(4), 1644-1654.

Healy, T. (2012). The unanticipated consequences of technology. *Nanotechnology: ethical and social Implications*, 155-173.

Hebinck, A., Vervoort, J. M., Hebinck, P., Rutting, L., and Galli, F. (2018). Imagining transformative futures: participatory foresight for food systems change. *Ecology and Society*, *23*(2).

Henderson, G. M. (2002). Transformative learning as a condition for transformational change in organizations. *Human Resource Development Review*, *1*(2), 186-214.

Henríquez, B. L. P. (2014). The problem of climate change: Challenges and opportunities in carbon governance. *Carbon Governance, Climate Change and Business Transformation*, 27-54.

Hermann, R. R., Mosgaard, M., and Kerndrup, S. (2016). The function of intermediaries in collaborative innovation processes: Retrofitting a Danish small island ferry with green technology. *International Journal of Innovation and Sustainable Development*, *10*(4), 361-383.

Herrfahrdt-Pähle, E., Schlüter, M., Olsson, P., Folke, C., Gelcich, S., and Pahl-Wostl, C. (2020). Sustainability transformations: socio-political shocks as opportunities for governance transitions. *Global Environmental Change*, *63*, 102097.

Herrmann, C., Cerdas, F., Abraham, T., Büth, L., and Mennenga, M. (2021). Biological transformation of manufacturing as a pathway towards environmental sustainability: Calling for systemic thinking. *CIRP Journal of Manufacturing Science and Technology*, *34*, 157-170.

Heucher, K., Alt, E., Soderstrom, S., Scully, M., and Glavas, A. (2024). Catalyzing action on social and environmental challenges: An integrative review of insider social change agents. *Academy of Management Annals*, *18*(1), 295-347.

Hickmann, T., Widerberg, O., Lederer, M., and Pattberg, P. (2021). The United Nations Framework Convention on Climate Change Secretariat as an orchestrator in global climate policymaking. *International Review of Administrative Sciences*, *87*(1), 21-38.

Hidalgo, D., Martín-Marroquín, J. M., and Corona, F. (2019). A multi-waste management concept as a basis towards a circular economy model. *Renewable and Sustainable Energy Reviews*, *111*, 481-489.

Higgott, R. (1993). Economic cooperation: theoretical opportunities and practical constraints. *The Pacific Review*, *6*(2), 103-117.

Hillerbrand, R. (2010). Unintended Consequences and Risky Technologies: A Virtue Ethical Approach to the Moral Problems Caused by Genetic Engineering. In *Managing Environmental Justice* (pp. 165-183*). Brill.*

Himeur, Y., Elnour, M., Fadli, F., Meskin, N., Petri, I., Rezgui, Y., ... and Amira, A. (2023). AI-big data analytics for building automation and management systems: a survey, actual challenges and future perspectives. *Artificial Intelligence Review*, *56*(6), 4929-5021.

Himeur, Y., Rimal, B., Tiwary, A., and Amira, A. (2022). Using artificial intelligence and data fusion for environmental monitoring: A review and future perspectives. *Information Fusion*, *86*, 44-75.

Himeur, Y., Rimal, B., Tiwary, A., and Amira, A. (2022). Using artificial intelligence and data fusion for environmental monitoring: A review and future perspectives. *Information Fusion*, *86*, 44-75.

Ho, P. (2001). Greening without conflict? Environmentalism, NGOs and civil society in China. *Development and Change*, *32*(5), 893-921.

Hodson, M., and Marvin, S. (2017). Intensifying or transforming sustainable cities? Fragmented logics of urban environmentalism. *Local Environment*, *22*(sup1), 8-22.

Hoffman, M., Schenck, C. J., and Herbst, F. (2022). Exploring the intersection where business models, a circular economy and sustainability meet in the waste economy: A scoping review. *Sustainability*, *14*(6), 3687.

Hoicka, C. E., Conroy, J., and Berka, A. L. (2021). Reconfiguring actors and infrastructure in city renewable energy transitions: A regional perspective. *Energy Policy*, *158*, 112544.

Holdgate, M. (1994). Environmental challenges and priorities for action. *Marine Pollution Bulletin*, *29*(6-12), 258-261.

Holland, B. (2008). Ecology and the limits of justice: Establishing capability ceilings in Nussbaum's capabilities approach. *Journal of Human Development*, *9*(3), 401-425.

Holland, B. (2008). Justice and the environment in Nussbaum's "Capabilities Approach" why sustainable ecological capacity is a meta-capability. *Political Research Quarterly*, *61*(2), 319-332.

Hong, M., and Chen, E. Y. X. (2017). Chemically recyclable polymers: a circular economy approach to sustainability. *Green Chemistry*, *19*(16), 3692-3706.

Horan, D. (2019). A new approach to partnerships for SDG transformations. *Sustainability*, *11*(18), 4947.

Horev, T., Pesis-Katz, I., and Mukamel, D. B. (2004). Trends in geographic disparities in allocation of health care resources in the US. *Health policy*, *68*(2), 223-232.

Howes, M. (1979). The uses of indigenous technical knowledge in development. *The IDS Bulletin*, *10*(2), 12-23.

Hsu, Y. C., Verma, H., Mauri, A., Nourbakhsh, I., and Bozzon, A. (2022). Empowering local communities using artificial intelligence. *Patterns*, *3*(3).

Hu, H., Yang, Q., Lu, X., Wang, W., Wang, S., and Fan, M. (2010). Air pollution and control in different areas of China. *Critical Reviews in Environmental Science and Technology*, *40*(6), 452-518.

Hua, S., Jin, S., and Jiang, S. (2024). The limitations and ethical considerations of ChatGPT. *Data intelligence*, *6*(1), 201-239.

Huang, L., Wu, J., and Yan, L. (2015). Defining and measuring urban sustainability: a review of indicators. *Landscape ecology*, *30*, 1175-1193.

Hughes-Hassell, S. (2020). Collection management for youth: Equity, inclusion, and learning. *American Library Association.*

Hung, L., Mann, J., Perry, J., Berndt, A., and Wong, J. (2022). Technological risks and ethical implications of using robots in long-term care. *Journal of Rehabilitation and Assistive Technologies Engineering*, *9*, 20556683221106917.

Hunter Jr, M. L., and Gibbs, J. P. (2006). *Fundamentals of conservation biology*. John Wiley and Sons.

Hurlimann, A. C., Moosavi, S., and Browne, G. R. (2021). Climate change transformation: A definition and typology to guide decision making in urban environments. *Sustainable Cities and Society*, *70*, 102890.

Hussain, M., Tapinos, E., and Knight, L. (2017). Scenario-driven roadmapping for technology foresight. *Technological Forecasting and Social Change*, *124*, 160-177.

Ifejika Speranza, C., Kiteme, B., Ambenje, P., Wiesmann, U., and Makali, S. (2010). Indigenous knowledge related to climate variability and change: insights from droughts in semi-arid areas of former Makueni District, Kenya. *Climatic change*, *100*(2), 295-315.

Ignatow, G. (2008). Globalization and the Transformation of Environmental Activism: Turkey since the 1980s. *Globalizations*, *5*(3), 433-447.

Inoue, C. Y. A., and Franchini, M. (2020). Socio-environmentalism. In *International Relations from the Global South* (pp. 296-314). *Routledge*

Ioannidis, A., Papandreou, A., and Sartzetakis, E. (2000). International environmental agreements: A literature review. *Cahiers de Recherche du GREEN 00-08.* Québec: Université Laval.

Isaac, G., Finn, S., Joe, J. R., Hoover, E., Gone, J. P., Lefthand-Begay, C., and Hill, S. (2018). Native American perspectives on health and traditional ecological knowledge. *Environmental health perspectives*, *126*(12), 125002.

Issaka, S., and Ashraf, M. A. (2017). Impact of soil erosion and degradation on water quality: a review. *Geology, Ecology, and Landscapes*, *1*(1), 1-11.

Jackson, A. (2014). Evaluation of public policy advocacy: Challenges, principles and BEST-AC case study. *International Journal of Public Sector Management, 27(4), 272-280.*

Jackson, T. (2014). Sustainable consumption. In *Handbook of sustainable development* (pp. 279-290). Edward Elgar Publishing.

Jakab, Z., Selbie, D., Squires, N., Mustafa, S., and Saikat, S. (2021). Building the evidence base for global health policy: the need to strengthen institutional networks, geographical representation, and global collaboration. *BMJ Global Health*, *6*(8), e006852.

Jalbout, M., and Farah, S. (2016). *Will the technology disruption widen or close the skills gap in the Middle East and North Africa?* Washington, DC: Global Economy and Development at Brookings Institute.

Jamison, A. (1996). The shaping of the global environmental agenda: the role of non-governmental organizations. *Risk, Environment, and Modernity. Towards a New Ecology,* London, Sage, 224-245.

Jamison, A. (2001). The making of green knowledge: Environmental politics and cultural transformation. *Cambridge University Press.*

Jangra, B., Majra, J. P., and Singh, M. (2016). Swachh Bharat Abhiyan (clean India mission): SWOT analysis. *International Journal of Community Medicine and Public Health, 3*(12), 3285-3290.

Jansen, K., and Vellema, S. (2008). One | Agribusiness and environmentalism: the politics of technology innovation and regulation. *Critical praise for this book*, 1.

Jansen, K., and Vellema, S. (Eds.). (2004). *Agribusiness and Society: Corporate responses to environmentalism, market opportunities, and public regulation*. Zed Books.

Janssen, W., Bouwman, H., van Buuren, R., and Haaker, T. (2014). An organizational competence model for innovation intermediaries. *European Journal of Innovation Management, 17*(1), 2-24.

Jaramillo, A., and Mendoza-Ponce, A. (2022). Climate change overview. In *The Impact of Climate Change on Fungal Diseases* (pp. 1-18). Cham: Springer International Publishing.

Javed, S., and Chattu, V. K. (2020). Strengthening the COVID-19 pandemic response, global leadership, and international cooperation through global health diplomacy. *Health promotion perspectives, 10*(4), 300.

Jewell, B., Blackman, A., Kuilboer, A., Hyvonen, T., Moscardo, G., and Foster, F. (2004). Factors contributing to successful tourism development in peripheral regions. *Journal of Tourism Studies, 15*(1), 59-70.

Jha, P. (2018). Poverty and Environmental Degradation in India. *Supremo Amicus, 8*, 168.

Jin, Y., Andersson, H., and Zhang, S. (2016). Air pollution control policies in China: a retrospective and prospects. *International journal of environmental research and public health, 13*(12), 1219.

Johnson, E. W., and Greenberg, P. (2017). The US environmental movement of the 1960s and 1970s: Building sustainability frameworks. In *Routledge Handbook of the History of Sustainability* (pp. 137-150). Routledge.

Johnson, J. T., Howitt, R., Cajete, G., Berkes, F., Louis, R. P., and Kliskey, A. (2016). Weaving Indigenous and sustainability sciences to diversify our methods. *Sustainability Science, 11*, 1-11.

Jolly, M. K., Preca, B. T., Tripathi, S. C., Jia, D., George, J. T., Hanash, S. M., ... and Levine, H. (2018). Interconnected feedback loops among ESRP1, HAS2, and CD44 regulate epithelial-mesenchymal plasticity in cancer. *APL bioengineering*, 2(3).

Jones, S. A., Moore, L. V., Moore, K., Zagorski, M., Brines, S. J., Roux, A. V. D., and Evenson, K. R. (2015). Disparities in physical activity resource availability in six US regions. *Preventive medicine, 78*, 17-22.

Jumaev, M. (2024). Leveraging Ai for Environmental Protection In Uzbekistan. *Journal of Scientific Research, Modern Views and Innovations, 1*(1), 140-144.

Kabra, G., Ramesh, A., and Arshinder, K. (2015). Identification and prioritization of coordination barriers in humanitarian supply chain management. *International Journal of Disaster Risk Reduction*, *13*, 128-138.

Kallis, G., and Bliss, S. (2019). Post-environmentalism: origins and evolution of a strange idea. *Journal of Political Ecology*, *26*(1), 466-485.

Karthikeyan, S., Ajay, G. B., Ahamed, N. R., and Sharun, A. (2022). *Edge AI–Based Aerial Monitoring. In Applied Edge AI* (pp. 237-260). Auerbach Publications.

Karunamuni, N., Imayama, I., and Goonetilleke, D. (2021). Pathways to well-being: Untangling the causal relationships among biopsychosocial variables. *Social science and medicine*, *272*, 112846.

Karwowski, M., and Raulinajtys-Grzybek, M. (2021). The application of corporate social responsibility (CSR) actions for mitigation of environmental, social, corporate governance (ESG) and reputational risk in integrated reports. *Corporate Social Responsibility and Environmental Management*, *28*(4), 1270-1284.

Kashif, M., Muaaz, S. H., Nafees, I., Anwar, R., and Ilma. (2023). Assessments for Disaster Risk Reduction by Analysing the Gaps and Future Scopes in Resilient Indigenous Practices. In *Indigenous Knowledge and Disaster Risk Reduction: Insight Towards Perception, Response, Adaptation and Sustainability* (pp. 417-439). Cham: Springer International Publishing.

Kasih, D. P. D., Dharmawan, N. K. S., Putra, I. B. W., Sudiarawan, K. A., and Rakhima, A. S. (2021). The Exploitation of Indigenous Communities by Commercial Actors. *Journal of Ethnic and Cultural Studies*, *8*(4), 91-108.

Kassas, M. (1997). Environment and social transformation. *Environmentalist*, *17*(1), 63-67.

Kayat, K., and Zainuddin, N. F. A. (2016). Community-based tourism initiative in rural Malaysia: Is it a success?. *International Review of Management and Marketing*, *6*(7), 242-249.

Kemfert, C., Breyer, C., and Oei, P. Y. (Eds.). (2020). 100% Renewable Energy Transition: Pathways and Implementation. *MDPI.*

Kemp, D. R., and Michalk, D. L. (2007). Towards sustainable grassland and livestock management. *The Journal of Agricultural Science*, *145*(6), 543-564.

Kendal, E. (2022). Ethical, legal and social implications of emerging technology (ELSIET) symposium. *Journal of Bioethical Inquiry*, *19*(3), 363-370.

Kent, M. (1987). Island biogeography and habitat conservation. *Progress in Physical Geography*, *11*(1), 91-102.

Kettl, D. F. (2000). The transformation of governance: Globalization, devolution, and the role of government. *Public Administration Review*, *60*(6), 488-497.

Kettl, D. F. (2000). The transformation of governance: Globalization, devolution, and the role of government. *Public Administration Review*, *60*(6), 488-497.

Kezar, A. (2013). Understanding sensemaking/sense giving in transformational change processes from the bottom up. *Higher education*, *65*, 761-780.

Khan, I., and Hou, F. (2021). Does multilateral environmental diplomacy improve environmental quality? The case of the United States. *Environmental Science and Pollution Research*, *28*(18), 23310-23322.

Khan, S. A. R., Umar, M., Asadov, A., Tanveer, M., and Yu, Z. (2022). Technological revolution and circular economy practices: a mechanism of green economy. *Sustainability*, *14*(8), 4524.

Khanzode, K. C. A., and Sarode, R. D. (2020). Advantages and disadvantages of artificial intelligence and machine learning: A literature review. *International Journal of Library and Information Science (IJLIS)*, *9*(1), 3. Fostering Community Engagement and Action

Kheira, T. (2021). Financial technology prospects in the Middle East and Africa. *Journal of Economic Growth*, *4*(3), 14-25.

Khojasteh, D., Davani, E., Shamsipour, A., Haghani, M., and Glamore, W. (2022). Climate change and COVID-19: Interdisciplinary perspectives from two global crises. *Science of the Total Environment*, *844*, 157142.

Kiamba, J. M. (2009). Women and leadership positions: Social and cultural barriers to success. *Wagadu Volume 6 Journal of International Women's Studies Volume 10: 1*, 89.

Kilkelly, U., Logan, E., Kilkelly, U., and Logan, E. (2021). Raising Awareness. *National Independent Human Rights Institutions for Children: Protecting and Promoting Children's Rights*, 45-63.

Kim, R. E. (2013). The emergent network structure of the multilateral environmental agreement system. *Global Environmental Change*, *23*(5), 980-991.

Kirchherr, J., Reike, D., and Hekkert, M. (2017). Conceptualizing the circular economy: An analysis of 114 definitions. *Resources, conservation and recycling*, *127*, 221-232.

Kirchner, J. W. (1989). The Gaia hypothesis: Can it be tested?. *Reviews of Geophysics*, *27*(2), 223-235.

Kirchner, J. W. (2002). The Gaia hypothesis: fact, theory, and wishful thinking. *Climatic change*, *52*(4), 391-408.

Klenk, N., Fiume, A., Meehan, K., and Gibbes, C. (2017). Local knowledge in climate adaptation research: moving knowledge frameworks from extraction to co-production. *Wiley Interdisciplinary Reviews: Climate Change*, *8*(5), e475.

Klijn, J. A. (2004). Driving forces behind landscape transformation in Europe, from a conceptual approach to policy options. *Frontis*, 201-218.

Koetz, T., Farrell, K. N., and Bridgewater, P. (2012). Building better science-policy interfaces for international environmental governance: assessing potential within the Intergovernmental Platform for Biodiversity and Ecosystem Services. *International environmental agreements: politics, law and economics*, *12*, 1-21.

König, M., Loibl, W., Haas, W., and Kranzl, L. (2015). Shared-socio-economic pathways. *Economic evaluation of climate change impacts: development of a cross-sectoral framework and results for Austria*, 75-99.

Kopnina, H. (2012). The Lorax complex: Deep ecology, ecocentrism and exclusion. *Journal of Integrative Environmental Sciences*, *9*(4), 235-254.

Kopnina, H. (2020). *Anthropocentrism: Problem of human-centered ethics in sustainable development goals. In Life on land* (pp. 48-57). Cham: Springer International Publishing.

Korhonen, J., Honkasalo, A., and Seppälä, J. (2018). Circular economy: the concept and its limitations. *Ecological economics*, *143*, 37-46.

Kostiner, I. (2003). *Conflicted legalities: A cultural analysis of law and activism.* University of California, Berkeley.

Kourula, A. (2010). Corporate engagement with non-governmental organizations in different institutional contexts—A case study of a forest products company. *Journal of World Business*, *45*(4), 395-404.

Kramsch, C. (2008). Ecological perspectives on foreign language education. *Language teaching*, *41*(3), 389-408.

Krautkraemer, J. A. (1998). Nonrenewable resource scarcity. *Journal of Economic literature*, *36*(4), 2065-2107.

Krueger, J. (2013). The Basel Convention and the international trade in hazardous wastes. In *Yearbook of International Cooperation on Environment and Development 2001-02* (pp. 43-51). Routledge.

Krupa, J., Poudineh, R., and Harvey, L. D. (2019). Renewable electricity finance in the resource-rich countries of the Middle East and North Africa: A case study on the Gulf Cooperation Council. *Energy*, *166*, 1047-1062.

Kuenkel, P., Kuenkel, P., and Paul. (2019). Stewarding sustainability transformations. Berlin, Germany: Springer International Publishing.

Küfeoğlu, S. (2022). *Emerging technologies: value creation for sustainable development* (p. 512). Springer Nature.

Kuhlmann, W. (1996). Making the law more ecocentric: Responding to Leopold and conservation biology. *Duke Envtl. L. and Pol'y F., 7, 133.*

Kumar, S., Darshna, A., and Ranjan, D. (2023). *A review of literature on the integration of green energy and circular economy.* Heliyon.

Kummer, K. (1998). The Basel Convention: Ten years on. *Rev. Eur. Comp. and Int'l Envtl. L.*, *7*, 227.

Kummer, K. (1999). *International management of hazardous wastes: the Basel Convention and related legal rules.* Oxford University Press, USA.

Kuo, C. C. (2008). A novel string structure for economic dispatch problems with practical constraints. *Energy Conversion and Management*, *49*(12), 3571-3577.

Kuziemsky, C. E., Randell, R., and Borycki, E. M. (2016). Understanding unintended consequences and health information technology. *Yearbook of medical informatics*, *25*(01), 53-60.

Kwok, J. Y. C. (2000). Public Participation and Community Empowerment (Doctoral dissertation, University of Auckland).

Kyvik, S., and Reymert, I. (2017). Research collaboration in groups and networks: differences across academic fields. *Scientometrics*, *113*, 951-967.

Lahijanian, A. (2011). Public participation in environmental education centers. *International Journal of Environmental Research*, *5*(4), 951-960.

Lal, R. A. T. T. A. N. (2001). Soil degradation by erosion. *Land degradation and development*, *12*(6), 519-539.

Lallas, P. L. (2001). The Stockholm Convention on persistent organic pollutants. *American Journal of International Law*, *95*(3), 692-708.

Lam, D. P., Hinz, E., Lang, D. J., Tengö, M., von Wehrden, H., and Martín-López, B. (2020). Indigenous and local knowledge in sustainability transformations research: a literature review. *Ecology and Society*, *25*(1).

Lamperti, F., Mandel, A., Napoletano, M., Sapio, A., Roventini, A., Balint, T., and Khorenzhenko, I. (2019). Towards agent-based integrated assessment models: examples, challenges, and future developments. *Regional environmental change, 19*, 747-762.

Lampkin, N., Padel, S., and Foster, C. (2000). Organic farming. In *CAP regimes and the European countryside: prospects for integration between agricultural, regional and environmental policies.* (pp. 221-238). Wallingford UK: CABI Publishing.

Landau, D. (2005). Transformational change: The case of Nationalab. *Organization Development Journal, 23*(4), 61.

Lange, A., and Vogt, C. (2003). Cooperation in international environmental negotiations due to a preference for equity. *Journal of Public Economics, 87*(9-10), 2049-2067.

Langlois-Bertrand, S., Benhaddadi, M., Jegen, M., and Pineau, P. O. (2015). Political-institutional barriers to energy efficiency. *Energy Strategy Reviews, 8*, 30-38.

Larkins, M. L. (2024). Introduction: practicing diversity, equity, inclusion, and justice in environmental studies and sciences. *Journal of Environmental Studies and Sciences, 14*(3), 443-451.

Larsen, K., and Hojer, M. (2007). Technological innovation and transformation perspectives in environmental futures studies for transport and mobility. *International Journal of Foresight and Innovation Policy, 3*(1), 95-115.

Lassa, J. A. (2019). Negotiating institutional pathways for sustaining climate change resilience and risk governance in Indonesia. *Climate, 7(8), 95.*

Lauber, V., and Jacobsson, S. (2015). Lessons from Germany's Energiewende. *The Triple Challenge for Europe: Economic Development, Climate Change and Governance*, 173-203.

Laverack, G. (2006). Improving health outcomes through community empowerment: a review of the literature. *Journal of Health, Population and Nutrition*, 113-120.

Leach, M., Bloom, G., Ely, A., Nightingale, P., Scoones, I., Shah, E., and Smith, A. (2007). Understanding governance: pathways to sustainability. *The Institute of Development Studies and partner organizations.*

Lee, H., McJeon, H., Yu, S., Liu, Y., Kim, H., and Eom, J. (2024). Decarbonization Pathways for Korea's Industrial Sector Towards Its 2050 Carbon Neutrality Goal. *Journal of Cleaner Production*, 143749.

Leendertse, K., and Taylor, P. (2011). Capacity building and knowledge sharing. *Water Resources Planning and Management*, 274-291.

Lempa, M., Goodman, R. M., Rice, J., and Becker, A. B. (2008). Development of scales measuring the capacity of community-based initiatives. *Health education and behavior, 35*(3), 298-315.

Lent, T., and Wells, R. P. (1992). Corporate environmental management study shows shift from compliance to strategy. *Environmental Quality Management, 1*(4), 379-394.

Leonardi, E., and Torre, S. (2022). Marxism and ecology: An ongoing debate. In *Handbook of Critical Environmental Politics* (pp. 71-89). Edward Elgar Publishing.

Leong, C., Pan, S. L., Bahri, S., and Fauzi, A. (2019). Social media empowerment in social movements: power activation and power accrual in digital activism. *European Journal of Information Systems, 28*(2), 173-204.

Leopold, A. (1992). *The river of the mother of God: and other essays by Aldo Leopold.* Univ of Wisconsin Press.

Lerner, A. M., Koshurina, V., Chistanova, O., and Wheeler, A. (2017). Mitigating the risks of resource extraction for industrial actors and northern indigenous peoples. *Arctic Review on Law and Politics*, *8*, 23-51.

Lessmann, O., and Rauschmayer, F. (2013). Re-conceptualizing sustainable development on the basis of the capability approach: A model and its difficulties. *Journal of human development and capabilities*, *14*(1), 95-114.

Lessmann, O., and Rauschmayer, F. (Eds.). (2014). *The capability approach and sustainability*. London: Routledge.

Li, H. L. (2011). Environmentalism and social foundations of education. In *Handbook of Research in the Social Foundations of Education (pp. 283-291).* Routledge.

Li, H. X., Edwards, D. J., Hosseini, M. R., and Costin, G. P. (2020). A review on renewable energy transition in Australia: An updated depiction. *Journal of cleaner production*, *242*, 118475.

Li, W. (2006). Environmental governance: Issues and challenges. *Envtl. L. Rep. News and Analysis*, *36*, 10505.

Liberti, L., Breckenridge, A., Hoekman, J., Leufkens, H., Lumpkin, M., McAuslane, N., ... and Rägo, L. (2016). Accelerating access to new medicines: Current status of facilitated regulatory pathways used by emerging regulatory authorities. *Journal of Public Health Policy*, *37*, 315-333.

Lim, J. H., and Tang, S. Y. (2002). Democratization and environmental policymaking in Korea. *Governance*, *15*(4), 561-582.

Lim, M. (2013). Many clicks but little sticks: Social media activism in Indonesia. *Digital activism in Asia reader*, 127-154.

Lin, B., and Bai, R. (2023). Nexus between green finance development and green technological innovation: A potential way to achieve the renewable energy transition. *Renewable Energy*, *218*, 119295.

Linares, I. M. P., De Paulo, A. F., and Porto, G. S. (2019). Patent-based network analysis to understand technological innovation pathways and trends. *Technology in society*, *59*, 101134.

Lindahl, K. B., Sandström, C., and Sténs, A. (2017). Alternative pathways to sustainability? Comparing forest governance models. *Forest Policy and Economics, 77, 69-78.*

Linnér, B. O., and Wibeck, V. (2019). *Sustainability transformations across societies: Agents and drivers across societies.* Cambridge University Press.

Linnér, B. O., and Wibeck, V. (2021). Drivers of sustainability transformations: leverage points, contexts and conjunctures. *Sustainability science*, *16*(3), 889-900.

Liporacci, H. S. N., Miranda, T. M., Hanazaki, N., and Peroni, N. (2015). How are legal matters related to the access of traditional knowledge being considered in the scope of ethnobotany publications in Brazil?. *Acta Botanica Brasilica*, *29*(2), 251-261.

Litvinenko, A., Lozina, J., and Chernikova, L. (2021). Imperfect Mechanisms for the Use and Protection of Intellectual Property as an Organizational Barrier in the Movement to Global Technological Leadership. *Technology and Business Strategy: Digital Uncertainty and Digital Solutions*, 211-223.

Liu, J., Yang, H., Gosling, S. N., Kummu, M., Flörke, M., Pfister, S., ... and Oki, T. (2017). Water scarcity assessments in the past, present, and future. *Earth's future*, *5*(6), 545-559.

Lo, C. F. (2009). Environmental Protection through FTAs: Paradigm Shifting from Multilateral to Multi-Bilateral Approach. *Asian J. WTO and Int'l Health L and Pol'y*, *4*, 309.

Loch, T. K., and Riechers, M. (2021). Integrating indigenous and local knowledge in management and research on coastal ecosystems in the Global South: A literature review. *Ocean and Coastal Management*, *212*, 105821.

Lockie, S., Fairley-Grenot, K., Ankeny, R., Botterill, L., Howlett, B., Mcbratney, A., ... and Woodhead, I. (2020). *The future of agricultural technologies*. Australian Council of Learned Academies (ACOLA).

Lofdahl, C. L. (2002). *Environmental impacts of globalization and trade: A systems study*. MIT Press.

Loikala, J., Hulkkonen, S., Itkyal, S., Kaushik, A., Keränen, S., Nisula, R., ... and Strand, S. (2006). Opportunities for Finnish environmental technology in India. *Sitra Reports, 63, 172.*

Loiseau, E., Saikku, L., Antikainen, R., Droste, N., Hansjürgens, B., Pitkänen, K., ... and Thomsen, M. (2016). Green economy and related concepts: An overview. *Journal of Cleaner Production*, *139*, 361-371.

Lomborg, B. (Ed.). (2004). Global crises, global solutions. *Cambridge university press.*

Lonergan, S. (1998). The Role of Environmental Degradation in. *Environmental change and security project report*, (4), 4.

Long, R. D., Charles, A., and Stephenson, R. L. (2015). Key principles of marine ecosystem-based management. *Marine Policy*, *57*, 53-60.

Loskutova, M. (2020). Quantifying scarcity: deforestation in the Upper Volga region and early debates over climate change in nineteenth-century Russia. *European Review of History: Revue européenne d'histoire*, *27*(3), 253-272.

Lozano, R. (2022). Toward sustainable organizations: A holistic perspective on implementation efforts. Springer Nature.

Lu, F., Murai, E., Campbell, S., and Angelo, H. (2024). Building more epistemically inclusive and environmentally equitable universities. *Journal of Environmental Studies and Sciences*, 1-14.

Ludovico de Almeida, M. F., and Caldas de Moraes, C. A. (2013). Diffusion of emerging technologies for sustainable development: Prospective assessment for public policies. *Journal of technology management and innovation*, *8*, 20-20.

Luintel, H., Bluffstone, R. A., and Scheller, R. M. (2018). The effects of the Nepal community forestry program on biodiversity conservation and carbon storage. *PLoS One*, *13*(6), e0199526.

Luintel, H., Bluffstone, R. A., Scheller, R. M., and Adhikari, B. (2017). The effect of the Nepal community forestry program on equity in benefit sharing. *The Journal of Environment and Development*, *26*(3), 297-321.

Luke, T. W. (2005). Environmentalism as globalization from above and below: Can world watchers truly represent the earth?. *Confronting globalization: Humanity, justice and the renewal of politics*, 154-171.

Luo, Y. (2024). Paradigm shift and theoretical implications for the era of global disorder. *Journal of International Business Studies*, *55*(2), 127-135.

Luttenberger, L. R. (2020). Waste management challenges in transition to a circular economy: the case of Croatia. *Journal of Cleaner Production*, *256*, 120495.

Lytle, M. H. (2007). *The gentle subversive: Rachel Carson, Silent Spring, and the rise of the environmental movement*. Oxford University Press.

Ma, T., and Wang, Y. (2021). Globalization and environment: Effects of international trade on emission intensity reduction of pollutants causing global and local concerns. *Journal of Environmental Management, 297, 113249.*

Maani, K. (2016). Multi-stakeholder decision making for complex problems: a systems thinking approach with cases. *World Scientific.*

Mabudafhasi, R. (2002). The role of knowledge management and information sharing in capacity building for sustainable development—an example from South Africa. *Ocean and Coastal Management*, *45*(9-10), 695-707.

Mace, G., Schreckenberg, K., and Poudyal, M. (2018). Ecosystem Services for Human Wellbeing. *Ecosystem Services And Poverty Alleviation*, 305.

Maggioni, E. (2015). Water demand management in times of drought: What matters for water conservation. *Water Resources Research*, *51*(1), 125-139.

Magliocca, N. R. (2023). Intersecting security, equity, and sustainability for transformation in the Anthropocene. *Anthropocene*, 100396.

Magnusdottir, G. L., and Kronsell, A. (2021). *Gender, intersectionality and climate institutions in industrialised states* (p. 278). Taylor and Francis.

Maiba, H. (2005). Grassroots transnational social movement activism: The case of Peoples' Global Action. *Sociological Focus*, *38*(1), 41-63.

Maina-Okori, N. M., Koushik, J. R., and Wilson, A. (2018). Reimagining intersectionality in environmental and sustainability education: A critical literature review. *The Journal of Environmental Education, 49(4), 286-296.*

Malin, S. A., Ryder, S., and Lyra, M. G. (2019). Environmental justice and natural resource extraction: Intersections of power, equity and access. *Environmental Sociology*, 5(2), 109-116.

Mansour, H. (2022). The Impediments to Advancing Artificial Intelligence in Healthcare within the Middle Eastern Context: A Critical Examination. *Journal of Advanced Analytics in Healthcare Management*, *6*(1), 58-65.

Marcinkowski, T. J. (2009). Contemporary challenges and opportunities in environmental education: Where are we headed and what deserves our attention?. *The journal of environmental education*, *41*(1), 34-54.

Maroyi, A. (2009). Traditional home gardens and rural livelihoods in Nhema, Zimbabwe: a sustainable agroforestry system. *International Journal of Sustainable Development and World Ecology*, *16*(1), 1-8.

Marquardt, J. (2016). *How power shapes energy transitions in Southeast Asia: A complex governance challenge*. Routledge.

Marshall, K. P. (1999). Has technology introduced new ethical problems?. *Journal of business ethics*, *19*, 81-90.

Martin, G. (2015). *Understanding social movements*. Routledge.

Martínez-Alier, J. (2020). A global environmental justice movement: mapping ecological distribution conflicts. *Disjuntiva. Crítica de Les Ciències Socials, 1*(2), 81-126.

Martinez-Alier, J., Temper, L., Del Bene, D., and Scheidel, A. (2016). Is there a global environmental justice movement?. *The Journal of Peasant Studies, 43*(3), 731-755.

Masten, A. S. (2018). Resilience theory and research on children and families: Past, present, and promise. *Journal of Family Theory and Review, 10*(1), 12-31.

Mati, B., and Penning de Vries, F. W. (2005). Bright Spots on technology-driven change in smallholder irrigation: case studies in Kenya. *Bright Spots Demonstrate Community Successes in African Agriculture*, 27-48.

Matthews, T. J., and Triantis, K. (2021). Island biogeography. *Current Biology, 31*(19), R1201-R1207.

Matz-Lück, N. (2012). Framework agreements. *Max Planck Encyclopedia of Public International Law, 4*, 220-24.

Mazurkiewicz, A. (2017, September). Technology transfer barriers encountered by RandD organisations. In *12th European Conference on Innovation and Entrepreneurship ECIE 2017* (p. 424).

Mazurkiewicz, A. (2018). Barriers of technological innovation development and implementation encountered by R&D organisations. *Journal of Machine Construction and Maintenance-Problemy Eksploatacji.*

Mbow, C. (2017). The Great Green Wall in the Sahel. In *Oxford Research Encyclopedia of Climate Science.*

McBeath, J., and Rosenberg, J. (2006). Comparative environmental politics (Vol. 25). Springer Science and Business Media.

McDermott, C. L., Montana, J., Bennett, A., Gueiros, C., Hamilton, R., Hirons, M., ... and Picot, L. (2023). Transforming land use governance: Global targets without equity miss the mark. *Environmental Policy and Governance, 33*(3), 245-257.

McGrail, S. (2011). Environmentalism in transition? Emerging perspectives, issues and futures practices in contemporary environmentalism. *Journal of Futures Studies, 15*(3), 117-144.

McIntosh, R. P. (1981). Succession and ecological theory. In *Forest succession: Concepts and application (pp. 10-23).* New York, NY: Springer, New York.

McKenzie-Mohr, D. (2000). New ways to promote pro-environmental behavior: Promoting sustainable behavior: An introduction to community-based social marketing. *Journal of Social Issues, 56*(3), 543-554.

McLaren, L., and Hawe, P. (2005). Ecological perspectives in health research. *Journal of Epidemiology and Community Health, 59*(1), 6-14.

McLeod, K. L., and Leslie, H. M. (2009). Why ecosystem-based management. *Ecosystem-based management for the oceans*, 3-12.

McNab, D., McKay, J., Shorrock, S., Luty, S., and Bowie, P. (2020). Development and application of 'systems thinking'principles for quality improvement. *BMJ open quality, 9*(1), e000714.

Mead, A., Griffiths, C. L., Branch, G. M., McQuaid, C. D., Blamey, L. K., Bolton, J. J., ... and Adams, J. B. (2013). Human-mediated drivers of change—impacts on coastal ecosystems and marine biota of South Africa. *African Journal of Marine Science, 35*(3), 403-425.

Mehmood, S., Zaman, K., Khan, S., and Ali, Z. (2024). The role of green industrial transformation in mitigating carbon emissions: Exploring the channels of technological innovation and environmental regulation. *Energy and Built Environment*, *5*(3), 464-479.

Melece, L. (2016). Challenges and opportunities of circular economy and green economy. *Engineering for Rural Development*, *25*, 1162-1169.

Mell, I. C. (2009, March). Can green infrastructure promote urban sustainability?. In *Proceedings of the Institution of Civil Engineers-engineering Sustainability* (Vol. 162, No. 1, pp. 23-34). Thomas Telford Ltd.

Menton, M., Larrea, C., Latorre, S., Martinez-Alier, J., Peck, M., Temper, L., and Walter, M. (2020). Environmental justice and the SDGs: from synergies to gaps and contradictions. *Sustainability science*, *15*, 1621-1636.

Mertig, A. G., and Dunlap, R. E. (2001). Environmentalism, new social movements, and the new class: A cross-national investigation. *Rural Sociology*, *66*(1), 113-136.

Meyer, J. M. (2001). *Political nature: Environmentalism and the interpretation of western thought.* Mit Press.

Michael, T. W. (2019). Sociocultural barriers to women's political leadership in the North Region of Cameroon. *Asian Journal of Humanities and Social Studies*, *7*(2).

Mihaylov, N. L., and Perkins, D. D. (2015). Local environmental grassroots activism: contributions from environmental psychology, sociology and politics. *Behavioral Sciences*, *5*(1), 121-153.

Miller, H. E., and Engemann, K. J. (2019). The precautionary principle and unintended consequences. *Kybernetes*, *48*(2), 265-286.

Miller, N. (2008). *Environmental politics: Stakeholders, interests, and policymaking.* Routledge.

Mimura, N. (2013). Overview of climate change impacts. *Climate change and global sustainability.* New York: United Nations, 46-57.

Miranda, L. A. (2010). Indigenous people as international lawmakers. *U. Pa. J. Int'l L.*, *32*, 203.

Misleh, D., Dziumla, J., De La Garza, M., and Guenther, E. (2024). Sustainability against the logics of the state: Political and institutional barriers in the Chilean infrastructure sector. *Environmental Innovation and Societal Transitions*, *51*, 100842.

Misleh, D., Dziumla, J., De La Garza, M., and Guenther, E. (2024). Sustainability against the logics of the state: Political and institutional barriers in the Chilean infrastructure sector. *Environmental Innovation and Societal Transitions*, *51*, 100842.

Misra, N. N., Dixit, Y., Al-Mallahi, A., Bhullar, M. S., Upadhyay, R., and Martynenko, A. (2020). IoT, big data, and artificial intelligence in agriculture and food industry. *IEEE Internet of things Journal*, *9*(9), 6305-6324.

Mitchell, I. K., Ling, C., Krusekopf, C., and Kerr, S. (2015). Pathways toward whole community transformation: a case study on the role of school engagement and environmental education. *Environment, Development and Sustainability*, *17*, 279-298.

Mohai, P., Pellow, D., and Roberts, J. T. (2009). Environmental justice. *Annual review of environment and resources*, *34*(1), 405-430.

Mohamed, A. A., and Nageye, A. I. (2019). Relationship between environmental degradation, resource scarcity, and civil conflicts in Somalia. *Journal of Environmental Management and Tourism, 10*(3), 640-650.

Mol, A. P. (2000). Ecological modernization: industrial transformations and environmental reform. *The International Handbook of Environmental Sociology;* Redclift, M., Woodgate, G., Eds, 138-149.

Monjardino, J., Dias, L., Fortes, P., Tente, H., Ferreira, F., and Seixas, J. (2021). Carbon neutrality pathways effects on air pollutant emissions: the Portuguese case. *Atmosphere, 12*(3), 324.

Monkelbaan, J. (2019). *Governance for the sustainable development goals.* Singapura: Spinger.

Moody, O. O. (2016). WIPO and the reinforcement of the Nagoya Protocol: Towards effective implementation of an access and benefit sharing regime for the protection of traditional knowledge associated with genetic resources (Master's thesis, Queen's University (Canada).

Moon, J. Y., Lee, U., Blain-Moraes, S., and Mashour, G. A. (2015). *General relationship of global topology, local dynamics.*

Moore, M., and Hartley, J. (2010). Innovations in governance. In *The New Public Governance? (pp. 68-87).* Routledge.

Morão, A., and Visser, D. (2022). Creating impact by integrating the SDGs in a company strategy. In *E3S Web of Conferences* (Vol. 349, p. 13003*).* EDP Sciences.

Morrison, A., Pietrobelli, C., and Rabellotti, R. (2008). Global value chains and technological capabilities: a framework to study learning and innovation in developing countries. *Oxford Development Studies, 36*(1), 39-58.

Morseletto, P. (2020). Restorative and regenerative: Exploring the concepts in the circular economy. *Journal of Industrial Ecology, 24*(4), 763-773.

Moss, B. (2008). Water pollution by agriculture. *Philosophical Transactions of the Royal Society B: Biological Sciences, 363*(1491), 659-666.

Moustakas, K., and Loizidou, M. (2023). Effective waste management with emphasis on circular economy. *Environmental Science and Pollution Research, 30*(4), 8540-8547.

Mugambi, J. N. (2016). Africa: African heritage and ecological stewardship. In *Routledge Handbook of Religion and Ecology* (pp. 109-119). Routledge.

Müller, F. (1997). State-of-the-art in ecosystem theory. *Ecological modeling, 100*(1-3), 135-161.

Munasinghe, M. (2011). *Addressing sustainable development and climate change together using sustainomics.* Wiley Interdisciplinary Reviews: Climate Change, *2*(1), 7-18.

Munda, G. (1997). Environmental economics, ecological economics, and the concept of sustainable development. *Environmental values, 6*(2), 213-233.

Mwangi, H., and Njiir, M. (2021). Mapping Coalitions: Mapping out coalitions, collaborations, partnerships and networks for media and civil society in sub-Saharan Africa. *Fojo: Media Institute.*

NA, N. (2016). *US environmentalism since 1945: A brief history with documents.* Springer.

Nair, P. R., Viswanath, S., and Lubina, P. A. (2017). Cinderella agroforestry systems. *Agroforestry Systems, 91*, 901-917.

Nanda, V., and Pring, G. R. (2012). *International environmental law and policy for the 21st century* (Vol. 9). Martinus Nijhoff Publishers.

Nath, R., Luan, Y., Yang, W., Yang, C., Chen, W., Li, Q., and Cui, X. (2015). Changes in arable land demand for food in India and China: A potential threat to food security. *Sustainability*, *7*(5), 5371-5397.

National Academies of Sciences, Policy, Global Affairs, Technology for Sustainability Program, Committee on Pathways to Urban Sustainability, and Opportunities. (2016). Pathways to urban sustainability: challenges and opportunities for the United States. National Academies Press- https://www.nationalacademies.org/our-work/pathways-to-urban-sustainability-challenges-and-opportunities.

Navickaite, J., and Janiunaite, B. (2012). Barriers of school principal's transformational leadership in change process: case study of Lithuanian schools. *Social Sciences*, *77*(3), 52-64.

Nazir, L., and Sharifi, A. (2024). An analysis of barriers to the implementation of smart grid technology in Pakistan. *Renewable Energy*, *220*, 119661.

Nelson, D. R. (2011). Adaptation and resilience: responding to a changing climate. *Wiley Interdisciplinary Reviews: Climate Change*, *2*(1), 113-120.

Nesmith, A. A., Schmitz, C. L., Machado-Escudero, Y., Billiot, S., Forbes, R. A., Powers, M. C., ... and Lawrence, L. A. (2021). Pathways to Change: Community and Environmental Transformation. *The Intersection of Environmental Justice, Climate Change, Community, and the Ecology of Life*, 83-104.

Never, B. (2012). Knowledge Systems and Change in Climate Governance: Comparing India and South Africa 2007-2010 (Doctoral dissertation, Staats-und Universitätsbibliothek Hamburg Carl von Ossietzky).

Nguyen, L. K. N., Kumar, C., Jiang, B., and Zimmermann, N. (2023). Implementation of systems thinking in public policy: A systematic review. *Systems*, *11*(2), 64.

Nikkhah, H. A., and Redzuan, M. R. B. (2010). The role of NGOs in promoting empowerment for sustainable community development. *Journal of Human Ecology*, *30*(2), 85-92.

Nilsson, M., and Persson, Å. (2017). Policy note: Lessons from environmental policy integration for implementing the 2030 Agenda. *Environmental Science and Policy*, *78*, 36-39.

Nkomoki, W., Bavorová, M., and Banout, J. (2018). Adoption of sustainable agricultural practices and food security threats: Effects of land tenure in Zambia. *Land use policy*, *78*, 532-538.

Norton, B. E. (1998). The application of grazing management to increase sustainable livestock production. *Animal production in Australia*, *22*(1), 15-26.

Novelli, M., Morgan, N., and Nibigira, C. (2012). Tourism in a post-conflict situation of fragility. *Annals of Tourism Research*, *39*(3), 1446-1469.

Nwanekezie, K., Noble, B., and Poelzer, G. (2022). Strategic assessment for energy transitions: A case study of renewable energy development in Saskatchewan, Canada. *Environmental Impact Assessment Review*, *92*, 106688.

Nygren, A. (1998). Environment as discourse: searching for sustainable development. *Environmental Values*, *7*(2), 201-222.

O'Brien, K. (2012). Global environmental change II: From adaptation to deliberate transformation. *Progress in human geography, 36*(5), 667-676.

O'Brien, K., and Sygna, L. (2013). Responding to climate change: the three spheres of transformation. *Proceedings of transformation in a changing climate, 16*, 23.

O'Meara, R. M., and O'Meara, R. M. (2014). Intended and Unanticipated Consequences. *Governing Military Technologies in the 21st Century: Ethics and Operations*, 44-65.

O'Riordan, T. (2004). Beyond environmentalism: towards sustainability. In *Unifying Geography* (pp. 117-143). *Routledge.*

Obidzinski, K., Andriani, R., Komarudin, H., and Andrianto, A. (2012). Environmental and social impacts of oil palm plantations and their implications for biofuel production in Indonesia. *Ecology and Society, 17*(1).

O'Brien, T. (2013). Growing green democracy? Barriers to ecological modernization in democratizing states. *Environmental Policy and Governance, 23*(4), 247-258.

Ofélia de Queiroz, F. A., Morte, I. B. B., Borges, C. L., Morgado, C. R., and de Medeiros, J. L. (2024). Beyond clean and affordable transition pathways: A review of issues and strategies to sustainable energy supply. *International Journal of Electrical Power and Energy Systems, 155*, 109544.

Oganisjana, K., Eremina, Y., Gvatua, S., Kabwende, B. N., and Chukwu, O. J. (2017). Barriers to social innovation and ways of overcoming them in Latvia. *Journal of Systemics, Cybernetics and Informatics, 15*(5), 33-38.

Okereke, C. (2008). Equity norms in global environmental governance. *Global Environmental Politics, 8*(3), 25-50.

Olson, D. M., Dinerstein, E., and Wikramanayake, E. D. (2002). Conservation biology for the biodiversity crisis. *Conservation Biology, 16*(1), 1-3.

O'Neill, K., Weinthal, E., Marion Suiseeya, K. R., Bernstein, S., Cohn, A., Stone, M. W., and Cashore, B. (2013). Methods and global environmental governance. *Annual Review of Environment and Resources, 38*(1), 441-471.

O'Neill, R. V., and Reichle, D. E. (1979). Dimensions of ecosystem theory (No. CONF-790462-1). Oak Ridge National Lab. (ORNL), Oak Ridge, TN (United States).

Opiyo, B., and Kuruç, Ü. K. (2016). Mediated Digital Activism: A Critical Assessment of Opportunities, Promises and Problems of Social Media Uses in Contemporary Grassroots Movements. *Online Journal of Communication and Media Technologies, 6 (December 2016-Special Issue),* 121-148.

Ordonez-Ponce, E. (2021). The role of institutional context for sustainability cross-sector partnerships. An exploratory analysis of European cities. *Sustainability, 13*(17), 9497.

O'Riordan, T. (1991). The new environmentalism and sustainable development. *Science of the Total Environment, 108*(1-2), 5-15.

O'Riordan, T. (2013). The challenge for environmentalism. In *New models in geography* (pp. 77-*102).* Routledge.

Oró, E., Depoorter, V., Garcia, A., and Salom, J. (2015). Energy efficiency and renewable energy integration in data centres. Strategies and modelling review. *Renewable and Sustainable Energy Reviews, 42*, 429-445.

Orr, D. (2014). Systems thinking and the future of cities. *Solutions, 5*(1), 54-61.

Ott, C., and Kiteme, B. (2016). Concepts and practices for the democratisation of knowledge generation in research partnerships for sustainable development. *Evidence and Policy*, *12*(3), 405-430.

Ott, K. (2003). The case for strong sustainability. *Greifswald's environmental ethics*, 59-64.

Otto, E. C. (2012). *Green speculations: Science fiction and transformative environmentalism*. The Ohio State University Press.

Oweis, T., Prinz, D., and Hachum, A. (2001). Water harvesting. *Indigenous knowledge for the future of the drier environments. International Centre for Agricultural Research in the Dry Areas (ICARDA),* Aleppo, Syria. p, 40.

Owen, L. A., Pickering, K. T., and Pickering, K. T. (2006). *An introduction to global environmental issues. Routledge.*

Özgüner, H. (2011). Cultural differences in attitudes towards urban parks and green spaces. *Landscape Research*, *36*(5), 599-620.

Paben, J. M. Z. (2011). Approaches to Environmental Justice: A Case Study of One Community's Victory. *S. Cal. Rev. L. and Soc. Just.*, *20*, 235.

Padilla, E. (2017). What can developing countries gain from a green transformation?. *Green Industrial Policy*.

Pádua, J. A. (2012). Environmentalism in Brazil: a historical perspective. *A companion to global environmental history*, 453-473.

Palmer, C., McShane, K., and Sandler, R. (2014). Environmental ethics. *Annual Review of Environment and Resources*, *39*(1), 419-442.

Panayotou, T. (2000). *Globalization and environment.* CID Working Paper Series.

Paoli, G. D., Wells, P. L., Meijaard, E., Struebig, M. J., Marshall, A. J., Obidzinski, K., ... and d'Arcy, L. (2010). Biodiversity conservation in the REDD. *Carbon balance and management*, *5*, 1-9.

Papaioannou, T., Wield, D., and Chataway, J. (2009). Knowledge ecologies and ecosystems? An empirically grounded reflection on recent developments in innovation systems theory. *Environment and Planning C: Government and Policy*, *27*(2), 319-339.

Park, J., and Roome, N. (2017). *The ecology of the new economy: sustainable transformation of global information, communications and electronics industries*. Routledge.

Passmore, J. (2017). Environmentalism. *A companion to contemporary political philosophy*, 572-592.

Passmore, J. (2017). Environmentalism. *A companion to contemporary political philosophy*, 572-592.

Patten, B. C. (2014). Systems ecology and environmentalism: Getting the science right. Part I: Facets for a more holistic Nature Book of ecology. *Ecological modelling*, *293*, 4-21.

Patten, B. C. (2014). Systems ecology and environmentalism: Getting the science right. Part I: Facets for a more holistic Nature Book of ecology. *Ecological modelling*, *293*, 4-21.

Patterson, J., Schulz, K., Vervoort, J., Van Der Hel, S., Widerberg, O., Adler, C., ... and Barau, A. (2017). Exploring the governance and politics of transformations towards sustainability. *Environmental Innovation and Societal Transitions*, *24*, 1-16.

Pawluk, R. R., Sandor, J. A., and Tabor, J. A. (1992). The role of indigenous soil knowledge in agricultural development. *Journal of soil and water conservation*, *47*(4), 298-302.

Pazicni, S., and Flynn, A. B. (2019). Systems thinking in chemistry education: Theoretical challenges and opportunities. *Journal of chemical education*, *96*(12), 2752-2763.

Pearce, J. (2000). Development, NGOs, and civil society: the debate and its future. *Development, NGOs, and civil society*, 15-43.

Peet, R. K., and Christensen, N. L. (1980). Succession: a population process. In *Succession: Symposium on advances in vegetation sciences, Nijmegen, The Netherlands, May 1979* (pp. 131-140). Springer Netherlands.

Peet, R., and Watts, M. (2004). *Liberation ecologies: environment, development and social movements*. Routledge.

Peeters, W., Dirix, J., and Sterckx, S. (2015). The capabilities approach and environmental sustainability: the case for functioning constraints. *Environmental values*, *24*(3), 367-389.

Pena, M. (2017). The Role of Intermediaries and Third Parties in Victim Participation. Victim Participation in International Criminal Justice*: Practitioners' Guide, 81-109.*

Penna, C. C., and Geels, F. W. (2012). Multi-dimensional struggles in the greening of industry: A dialectic issue lifecycle model and case study. *Technological Forecasting and Social Change*, *79*(6), 999-1020.

Pepper, D. (2019). *The roots of modern environmentalism*. Routledge.

Pereira, A. R., and Laube, W. (2018). Knowledge politics in environmental conflicts: A case from Brazil. *Gestión y Ambiente*, *21*(2Supl), 30-46.

Pereira, R. H., Schwanen, T., and Banister, D. (2017). Distributive justice and equity in transportation. *Transport Reviews*, *37*(2), 170-191.

Perissi, I., and Jones, A. (2022). Investigating European Union decarbonization strategies: Evaluating the pathway to carbon neutrality by 2050. *Sustainability*, *14*(8), 4728.

Perkins, P. E. (2018). Climate justice, gender and intersectionality. In *Routledge handbook of climate justice* (pp. 349-358). Routledge.

Peterson, G. D., Cumming, G. S., and Carpenter, S. R. (2003). Scenario planning: a tool for conservation in an uncertain world. *Conservation biology, 17(2), 358-366.*

Petts, J. (2005). Enhancing environmental equity through decision-making: learning from waste management. *Local environment*, *10*(4), 397-409.

Pharino, C. (2019). *Paranormal tourism: Case studies in Southeast Asian countries* (Doctoral dissertation, James Cook University).

Pichler, M., Schaffartzik, A., Haberl, H., and Görg, C. (2017). Drivers of society-nature relations in the Anthropocene and their implications for sustainability transformations. *Current Opinion in Environmental Sustainability*, *26*, 32-36.

Pickering, J., Hickmann, T., Bäckstrand, K., Kalfagianni, A., Bloomfield, M., Mert, A., ... and Lo, A. Y. (2022). Democratising sustainability transformations: Assessing the transformative potential of democratic practices in environmental governance. *Earth System Governance*, *11*, 100131.

Pisano, U. (2012). Resilience and Sustainable Development: Theory of resilience, systems thinking. *European sustainable development network (ESDN), 26*, 50.

Planton, S., Déqué, M., Chauvin, F., and Terray, L. (2008). Expected impacts of climate change on extreme climate events. *Comptes Rendus Geoscience, 340(9-10), 564-574.*

Plater, Z. J. (2005). Dealing with Dumb and Dumber: The Continuing Mission of Citizen Environmentalism. *J. Envtl. L. and Litig., 20*, 9.

Polasky, S. (2006, July). Biodiversity Conservation and Ecosystem Services. In *Plenary Lecture, Third World Congress of Environmental and Resource Economists,* Kyoto.

Poncelet, E. C. (2001). Personal transformation in multistakeholder environmental partnerships. *Policy Sciences*, *34*(3), 273-301.

Porter-Bolland, L., Ellis, E. A., Guariguata, M. R., Ruiz-Mallén, I., Negrete-Yankelevich, S., and Reyes-García, V. (2012). Community managed forests and forest protected areas: An assessment of their conservation effectiveness across the tropics. *Forest ecology and management*, *268*, 6-17.

Preisendörfer, P. (2017). Environmental problems as drivers of economic and social transformation in Europe. In *The Routledge International Handbook of European Social Transformations* (pp. 212-223). *Routledge.*

Pringle, R., Michael, K., and Michael, M. G. (2016). Unintended consequences of living with AI: The paradox of technological potential? Part II [guest editorial]. *IEEE Technology and Society Magazine*, *35*(4), 17-21.

Prosser, I. P., Chiew, F. H., and Stafford Smith, M. (2021). Adapting water management to climate change in the Murray–Darling Basin, Australia. *Water*, *13*(18), 2504.

Pugh, J. (2008). Challenges and opportunities of human conflict and environmental transformation. *Pirineos*, *163*, 63-75.

Purcell, W. M., and Chahine, T. (2019). Leadership and governance frameworks driving transformational change in an entrepreneurial UK university. *Leadership and Organization Development Journal*, *40*(5), 612-623.

Qu, Z. (2014). Three Essays on the Role of Intermediaries (Doctoral dissertation, Christian-Albrechts Universität Kiel).

Quick, K. S., and Bryson, J. M. (2022). Public participation. In *Handbook on theories of governance* (pp. 158-168). Edward Elgar Publishing.

Radcliffe, S. A. (2012). Dismantling gaps and myths: How Indigenous political actors broke the mold of socioeconomic development. *The Brown Journal of World Affairs*, *18*(2), 89-102.

Radovanović, M., Filipović, S., and Andrejević Panić, A. (2021). Sustainable energy transition in Central Asia: status and challenges. *Energy, Sustainability and Society*, *11*, 1-13.

Rahman, M., Akter, M., Odunukan, K., and Haque, S. E. (2020). Examining economic and technology-related barriers of small-and medium-sized enterprises internationalisation: An emerging economy context. *Business Strategy and Development*, *3*(1), 16-27.

Raj, R., and Griffin, K. A. (Eds.). (2017). *Conflicts, religion, and culture in tourism.* CABI.

Rajak, S., Parthiban, P., and Dhanalakshmi, R. (2016). Sustainable transportation systems performance evaluation using fuzzy logic. *Ecological Indicators*, *71*, 503-513.

Ramadhani, D. (2024). The Role of Sensor Technology in the Development of Environmental Monitoring Systems: A Systematic Review of the Literature. *Science and Applied Journal (SIAP)*, *1*(1), 01-08.

Ramanathan, N. (2021). Quality-based management for future-ready corporations serving society and planet. *Total Quality Management and Business Excellence*, *32*(5-6), 541-557.

Rands, M. R., Adams, W. M., Bennun, L., Butchart, S. H., Clements, A., Coomes, D., ... and Vira, B. (2010). Biodiversity conservation: challenges beyond 2010. *Science*, *329*(5997), 1298-1303.

Rao, M., Nagendra, H., Shahabuddin, G., and Carrasco, L. R. (2016). Integrating community-managed areas into protected area systems: the promise of synergies and the reality of trade-offs. *Protected areas: are they safeguarding biodiversity?* 169-189.

Rask, N. (2022). An intersectional reading of circular economy policies: towards just and sufficiency-driven sustainability's. *Local Environment*, *27*(10-11), 1287-1303.

Rathi, A. K. A. (2021). *Handbook of environmental impact assessment: Concepts and practice*. Cambridge Scholars Publishing.

Reckien, D., Lwasa, S., Satterthwaite, D., McEvoy, D., Creutzig, F., Montgomery, M., ... and Bautista, E. (2018). Equity, environmental justice, and urban climate change. In *Climate change*

Redclift, M. (2005). Sustainable development (1987–2005): an oxymoron comes of age. *Sustainable development*, *13*(4), 212-227.

Reddy, B. S. (2010). Organic farming: status, issues and prospects–a review. *Agricultural Economics Research Review*, *23*(2), 343-358.

Reddy, Shaker P. C., and Sureshbabu, A. (2019). An Adaptive Model for Forecasting Seasonal Rainfall Using Predictive Analytics. *International Journal of Intelligent Engineering and Systems*, *12*(5).

Reed, B. M., Sarasan, V., Kane, M., Bunn, E., and Pence, V. C. (2011). Biodiversity conservation and conservation biotechnology tools. *In Vitro Cellular and Developmental Biology-Plant*, *47*, 1-4.

Reed, D. L. (2013). *Capability, environment, and deliberation: Applying the capabilities approach to environmental policy-making and institutional design*. University of California, Santa Barbara.

Reed, M. G., and George, C. (2011). Where in the world is environmental justice?. *Progress in Human Geography*, *35*(6), 835-842.

Rehman, A., Radulescu, M., Ma, H., Dagar, V., Hussain, I., and Khan, M. K. (2021). The impact of globalization, energy use, and trade on ecological footprint in Pakistan: does environmental sustainability exist?. *Energies, 14(17), 5234.*

Reid, A. (2019). Curriculum and environmental education: Perspectives, priorities and challenges. In *Curriculum and Environmental Education* (pp. 5-27). *Routledge.*

Rennings, K., Kemp, R., Bartolomeo, M., Hemmelskamp, J., and Hitchens, D. (2003). *Blueprints for an integration of science, technology and environmental policy (BLUEPRINT)*. Mannheim, Germany: Zentrum für Europäische Witschaftsführung GmbH (ZEW).

Reno, S. T. (2020). *Early Anthropocene Literature in Britain*, 1750–1884. Springer Nature.

Reo, N. J., Whyte, K. P., McGregor, D., Smith, M. A., and Jenkins, J. F. (2017). Factors that support Indigenous involvement in multi-actor environmental stewardship. *AlterNative: An International Journal of Indigenous Peoples*, *13*(2), 58-68.

Richardson, B. C. (1999). Toward a policy on a sustainable transportation system. *Transportation Research Record*, *1670*(1), 27-34.

Rios, S. M. (2018). *Valley Fever: Environmental Racism and Health Justice*. University of California, Santa Barbara.

Roberts, J. T. (2007). *Globalizing environmental justice*. MIT.

Roberts, J. T. (2008). Commentary: Challenges and opportunities for global environmental governance in the 21st century. *Global Environmental Change*, *18*(3), 375-379.

Robertson, T., Tucker, R. P., Breyfogle, N. B., and Mansoor, P. (Eds.). (2020). *Nature at war: American environments and World War II*. Cambridge University Press.

Robeyns, I. (2006). The capability approach in practice. *Journal of political philosophy*, *14*(3).

Robeyns, I. (2017). Assessing global poverty and inequality: income, resources, and capabilities. In *Justice and the Capabilities Approach* (pp. 411-430). *Routledge*.

Robinson, D. K., Huang, L., Guo, Y., and Porter, A. L. (2013). Forecasting Innovation Pathways (FIP) for new and emerging science and technologies. *Technological Forecasting and Social Change*, *80*(2), 267-285.

Robinson, J. G. (1993). The limits to caring: Sustainable living and the loss of biodiversity. *Conservation Biology*, *7*(1), 20–28. https://doi.org/10.1046/j.1523-1739.1993.07010020.x

Rock, M. T., and Angel, D. P. (2005). *Industrial transformation in the developing world*. Oxford University Press.

Rodriguez, I. (2020). Latin American decolonial environmental justice. In *Environmental Justice* (pp. 78-93). Routledge.

Rodriguez, P., and Costa, I. (2024). Exploring the Role of AI in Sustainable Development and Environmental Monitoring. *MZ Computing Journal*, *5*(1).

Roesler, S. M. (2011). Addressing environmental injustices: a capability approach to rulemaking. *W. Va. L. Rev.*, *114*, 49.

Rolston, H. (1988). *Environmental ethics*. Temple University Press.

Ronzhyn, A., and Wimmer, M. A. (2019). Literature review of ethical concerns in the use of disruptive technologies in government 3.0. *In Icds (pp. 85-93).*

Rootes, C. (2003). The transformation of environmental activism: An introduction. *Environmental Protest in Western Europe*, 1-19.

Rootes, C. A. (1999). The transformation of environmental activism: Activists, organizations and policy-making. *Innovation: The European Journal of Social Science Research*, *12*(2), 155-173.

Rosenbloom, D., and Meadowcroft, J. (2022). Accelerating pathways to net zero: Governance strategies from transition studies and the transition accelerator. *Current Climate Change Reports*, *8*(4), 104-114.

Rotillon, G., Tazdaït, T., and Zeghni, S. (1996). Bilateral or multilateral bargaining in the face of global environmental change?. *Ecological Economics*, *18*(2), 177-187.

Roux, J. L., Konczal, A. A., Bernasconi, A., Bhagwat, S. A., De Vreese, R., Doimo, I., ... and Winkel, G. (2022). Exploring evolving spiritual values of forests in Europe and

Asia: a transition hypothesis toward re-spiritualizing forests. *Ecology and Society*, *27*(4), 1-20.

Rowe, G., and Frewer, L. J. (2004). Evaluating public-participation exercises: a research agenda. *Science, technology, and human values*, *29*(4), 512-556.

Roy, M., and Venema, H. D. (2002). Reducing risk and vulnerability to climate change in India: the capabilities approach. *Gender and Development*, *10*(2), 78-83.

Roy, S., Rautela, R., and Kumar, S. (2023). Towards a sustainable future: Nexus between the sustainable development goals and waste management in the built environment. *Journal of Cleaner Production*, *415*, 137865.

Rudito, B., Famiola, M., and Anggahegari, P. (2022). Corporate social responsibility and social capital: journey of community engagement toward community empowerment program in developing country. *Sustainability*, *15*(1), 466.

Ruiz-Mallén, I., Schunko, C., Corbera, E., Rös, M., and Reyes-García, V. (2015). Meanings, drivers, and motivations for community-based conservation in Latin America. *Ecology and Society*, *20*(3).

Russell, K., Rosenbaum, S., Varela, S., Stanton, R., and Barnett, F. (2023). Fostering community engagement, participation and empowerment for mental health of adults living in rural communities: a systematic review. *Rural and remote health*, *23*(1), 1-13.

Ryder, S. S. (2017). A bridge to challenging environmental inequality: Intersectionality, environmental justice, and disaster vulnerability. *Social Thought and Research: A Continuation of the Mid-American Review of Sociology*, 85-115.

Ryff, C. D., Singer, B. H., and Seltzer, M. M. (2002). 12 Pathways through challenge: implications for well-being and health. Paths to successful development*: Personality in the life course, 302.*

Saarela, S. R. (2020). In between two worlds? Science-policy interaction in Finnish environmental governance (Doctoral dissertation, Doctoral dissertation. University of Helsinki, Faculty of Biological and Environmental Sciences. Doctoral Programme in Interdisciplinary Environmental Sciences. Helsingin yliopisto. http://urn. fi/URN).

Sachs, J. D., Schmidt-Traub, G., Mazzucato, M., Messner, D., Nakicenovic, N., and Rockström, J. (2019). Six transformations to achieve the sustainable development goals. *Nature sustainability*, *2*(9), 805-814.

Sagoff, M. (1995). Carrying capacity and ecological economics. *BioScience*, *45*(9), 610-620.

Salam, A. (2024). *Internet of Things for sustainable community development: wireless communications, sensing, and systems*. Springer Nature.

Saleem, F., Zhang-Zhang, Y., Malik, M. I., and Allui, A. (2020). Revisiting stakeholder theory and environmentalism: Evidence from an emerging economy. *Sustainability*, *12*(20), 8751.

Salmon, P. M., Stanton, N. A., Walker, G. H., Hulme, A., Goode, N., Thompson, J., and Read, G. J. (2022). *Handbook of systems thinking methods*. CRC Press.

Sands, P. (1992). The United Nations framework convention on climate change. *Rev. Eur. Comp. and Int'l Envtl. L.*, *1*, 270.

Saner, M. A. (2013). The role of adaptation in the governance of emerging technologies. In *Innovative Governance Models for Emerging Technologies* (pp. 92-107). Edward Elgar Publishing.

Sangha, K. K. (2020). Global importance of indigenous and local communities' managed lands: building a case for stewardship schemes. *Sustainability, 12*(19), 7839.

Sapiński, A., Sanetra-Półgrabi, S., Kasian, S. Y., Rozhko, N., and Bhandari, M. P. (2022). Social Responsibility as a Tool for the Human Resources Policy Development and Reducing Inequalities on Tourism Industry. In *Inequality–the unbeatable challenge* (pp. 293-306). River Publishers.

Sarkki, S., Balian, E., Heink, U., Keune, H., Nesshöver, C., Niemelä, J., ... and Young, J. C. (2020). Managing science-policy interfaces for impact: Interactions within the environmental governance meshwork. *Environmental Science and Policy, 113*, 21-30.

Sawin, J. L., and Zehr, H. (2013). The ideas of engagement and empowerment. In *Handbook of restorative justice* (pp. 41-58). Willan.

Schang, K. A., Trant, A. J., Bohnert, S. A., Closs, A. M., Humchitt, M., McIntosh, K. P., ... and Wickham, S. B. (2020). Ecological research should consider Indigenous peoples and stewardship. *Facets, 5*(1), 534-537.

Schilderman, T. (2004). Adapting traditional shelter for disaster mitigation and reconstruction: experiences with community-based approaches. *Building research and information, 32*(5), 414-426.

Schleicher, J., Schaafsma, M., and Vira, B. (2018). Will the Sustainable Development Goals address the links between poverty and the natural environment?. *Current Opinion in Environmental Sustainability, 34*, 43-47.

Schmitt, M., and Maes, J. (2013). Equity and justice. In *Psychology of entertainment* (pp. 273-289). Routledge.

Schofer, E., and Granados, F. J. (2006). Environmentalism, globalization and national economies, 1980–2000. *Social Forces, 85*(2), 965-991.

Schokkaert, E. (2008). The capabilities approach. *Available at SSRN 1084821*.

Scholte, J. A. (2018). Civil society and NGOs. In *International organization and global governance* (pp. 351-364). Routledge.

Schreurs, J., Martens, P., and Kok, G. (2012). Living with less as a transformation process: A qualitative study of consumer behavior reform through spending reduction. *Qualitative Market Research: An International Journal, 15*(2), 188-205.

Schroeder, C. C. (1999). Collaboration and partnerships. *Higher education trends for the next century: A research agenda for student success*, 44-49.

Seckler, D., Barker, R., and Amarasinghe, U. (1999). Water scarcity in the twenty-first century. *International Journal of Water Resources Development, 15*(1-2), 29-42.

Sen, S., and Azonobi, L. M. (2004). Environmental Justice in Transportation Planning and Policy: Some Evidence from Practice in the Baltimore-Washington DC Metropolitan Region. Baltimore, MD: National Transportation Center, Morgan State University.

Sessions, G. (1987). The deep ecology movement: A review. *Environmental review, 11*(2), 105-125.

Sessions, G. (1988). Ecocentrism and The Greens: Deep Ecology and the Enviromental Task. *The Trumpeter, 5*(2).

Seufert, V., Ramankutty, N., and Mayerhofer, T. (2017). What is this thing called organic?–How is organic farming codified in regulations? *Food Policy*, *68*, 10-20.

Shafiee, S., and Topal, E. (2009). When will fossil fuel reserves be diminished? *Energy policy*, *37*(1), 181-189.

Shakesprere, J., Mizota, M., Martinez, R., Daly, H., and Falkenburger, E. (2021). Fostering Partnerships for Community Engagement. *Washington, DC: Urban Institute*.

Shao, T., Pan, X., Li, X., Zhou, S., Zhang, S., and Chen, W. (2022). China's industrial decarbonization in the context of carbon neutrality: A sub-sectoral analysis based on integrated modelling. *Renewable and Sustainable Energy Reviews*, *170*, 112992.

Shao, Y., and Chen, Z. (2022). Can government subsidies promote the green technology innovation transformation? Evidence from Chinese listed companies. *Economic Analysis and Policy*, *74*, 716-727.

Shellenberger, M., and Nordhaus, T. (2009). The death of environmentalism. *Geopolitics, History, and International Relations*, *1*(1), 121-163.

Shelomentsev, A., Medvedev, A., Doroshenko, S., Kozlova, O., and Komotsky, E. (2016). An Assessment of Barriers in Regional Development. *International Multidisciplinary Scientific GeoConference: SGEM*, *3*, 11-18.

Shepard, P. M. (1993). Issues of community empowerment. *Fordham Urb. LJ*, *21*, 739.

Sherwood, D. (2011). *Seeing the forest for the trees: a manager's guide to applying systems thinking*. Nicholas Brealey International.

Shiva, V., and Mies, M. (2014). *Ecofeminism.* Bloomsbury Publishing.

Shkaruba, A., Skryhan, H., Likhacheva, O., Kireyeu, V., Katona, A., Shyrokostup, S., and Sepp, K. (2021). Environmental drivers and sustainable transition of dachas in Eastern Europe: An analytical overview. *Land use policy*, *100*, 104887.

Shrivastava, P., and Scott, H. I. (1992). Corporate self-greenewal: Strategic responses to environmentalism. *Business Strategy and the Environment*, *1*(3), 9-21.

Shrivastava, P., Smith, M. S., O'Brien, K., and Zsolnai, L. (2020). Transforming sustainability science to generate positive social and environmental change globally. *One Earth*, 2(4), 329-340.

Sidabalok, H. (2023). *Residential Solid Waste Management in Semarang: The Question of Geographical Environmental Justice. PT Kanisius.*

Simpson, M. C. (2008). Community benefit tourism initiatives—A conceptual oxymoron?. *Tourism Management*, *29*(1), 1-18.

Singh, R. L., and Singh, P. K. (2017). Global environmental problems. *Principles and environmental biotechnology applications for a sustainable future*, 13-41.

Singh, S. (2021). Energy crisis and climate change: Global concerns and their solutions. *Energy: crises, challenges and solutions*, 1-17.

Singleton, B. E., Rask, N., Magnusdottir, G. L., and Kronsell, A. (2022). Intersectionality and climate policy-making: The inclusion of social difference by three Swedish government agencies. *Environment and Planning C: Politics and Space*, *40*(1), 180-200.

Sirin, C. V. (2011). Scarcity-induced domestic conflict: examining the interactive effects of environmental scarcity and 'Ethnic'population pressures. *Civil Wars*, *13*(2), 122-140.

Sivanappan, R. K. (2006, November). Rainwater harvesting, conservation, and management strategies for urban and rural sectors. In *National seminar on rainwater harvesting and water management (Vol. 11, No. 12, p. 1)*. New Delhi, India: Institution of Engineers (India), Nagpur Local Centre, Nagpur in association with UNESCO.

Slepchuk, A. N., Milne, G. R., and Swani, K. (2022). Overcoming privacy concerns in consumers' use of health information technologies: A justice framework. *Journal of Business Research, 141*, 782-793.

Smil, V. (1987). A perspective on global environmental crises. *Futures, 19*(3), 240-253.

Smith, A., and Ely, A. (2015). Green Transformations From Below?: The politics of grassroots innovation. In *The politics of green transformations* (pp. 102-118). *Routledge.*

Smith, A., and Pulver, S. (2009). Ethics-based environmentalism in practice: Religious-environmental organizations in the United States. *Worldviews: Global Religions, Culture, and Ecology, 13*(2), 145-179.

Smith, A., Fressoli, M., and Thomas, H. (2014). Grassroots innovation movements: challenges and contributions. *Journal of Cleaner Production, 63*, 114-124.

Smith, B. M., Chakrabarti, P., Chatterjee, A., Chatterjee, S., Dey, U. K., Dicks, L. V., ... and Basu, P. (2017). Collating and validating indigenous and local knowledge to apply multiple knowledge systems to an environmental challenge: A case-study of pollinators in India. *Biological conservation, 211*, 20-28.

Smith, R. J., Muir, R. D., Walpole, M. J., Balmford, A., and Leader-Williams, N. (2003). Governance and the loss of biodiversity. *Nature, 426*(6962), 67-70.

Söderholm, P. (2020). The green economy transition: the challenges of technological change for sustainability. *Sustainable Earth, 3*(1), 6.

Söderholm, P. (2020). The green economy transition: the challenges of technological change for sustainability. *Sustainable Earth, 3*(1), 6.

Speed, C. (2006). Anthropocentrism and sustainable development: Oxymoron or symbiosis. *Transactions on Ecology and teh Environment, 93*, 323-332.

Speth, J. G., and Haas, P. M. (2007). Global environmental governance. *Pearson Education India.*

Stahel, W. R. (2016). The circular economy. *Nature, 531*(7595), 435-438.

Stammers, N., and Eschle, C. (2005). Social movements and global activism. *Global activism, global media*, 50-67.

Stefik, M. (2014). *Introduction to knowledge systems*. Elsevier.

Steinveg, B. (2023). Actors. In *Arctic Governance Through Conferencing: Actors, Agendas and Arenas* (pp. 79-106). Cham: Springer International Publishing.

Stenn, T. L. (2016). *Social entrepreneurship as sustainable development: Introducing the sustainability lens*. Springer.

Stephan, H. R., and Stephan, H. R. (2015). Environmental History: Nature, Landscapes, and Identities. *Cultural Politics and the Transatlantic Divide over GMOs*, 132-177.

Stephan, H. R., and Zelli, F. (2018). International Organizations and the Global Environment. In *Politics of the Environment* (pp. 52-70). *Routledge.*

Stevis, D., and Boswell, T. (2007). International Framework Agreements. *Global unions: Challenging transnational capital through cross-border campaigns*, 174.

Stewardson, M. J., Bond, N., Brookes, J., Capon, S., Dyer, F., Grace, M., ... and Western, A. W. (2021). The politicization of science in the Murray-Darling Basin, Australia: discussion of 'Scientific integrity, public policy, and water governance.' *Australasian Journal of Water Resources*, *25*(2), 141-158.

Stobart, J. (2013). The first industrial region: north-west England, c. 1700–60. In *The first industrial region*. Manchester University Press.

Stockdale, S., Patel, K., Gray, R., Hill, D. A., Franklin, C., and Madyun, N. (2006). Supporting Wellness Through Policy and Advocacy. *Ethnicity and Disease, 16, 43-53.*

Stojanovic, T., Green, D. R., and Lymbery, G. (2010). Approaches to knowledge sharing and capacity building: The role of local information systems in marine and coastal management. *Ocean and Coastal Management*, *53*(12), 805-815.

Stone, B., Vargo, J., and Habeeb, D. (2012). Managing climate change in cities: Will climate action plans work?. *Landscape and urban planning*, *107*(3), 263-271.

Su, X., Xu, A., Lin, W., Chen, Y., Liu, S., and Xu, W. (2020). Environmental leadership, green innovation practices, environmental knowledge learning, and firm performance. *Sage Open*, *10*(2), 2158244020922909.

Suchek, N., Fernandes, C. I., Kraus, S., Filser, M., and Sjögrén, H. (2021). Innovation and the circular economy: A systematic literature review. *Business Strategy and the Environment*, *30*(8), 3686-3702.

Sujee, S., and Solanki, R. (2024). Adoption and Proliferation of Digital Technology in the Banking Sector of the Middle East. In *Handbook of Banking and Finance in the MENA Region* (pp. 3-23).

Sulaiman, M. H., Daniyal, H., and Mustafa, M. W. (2012, December). Modified firefly algorithm in solving economic dispatch problems with practical constraints. In *2012 IEEE International Conference on Power and Energy (PECon)* (pp. 157-161*). IEEE.*

Sulecio de Alvarez, M., and Dickson-Deane, C. (2018). Avoiding educational technology pitfalls for inclusion and equity. *TechTrends*, *62*(4), 345-353.

Šupa, M., and Kruopštaitė, I. (2022). Deviance and ethical considerations in online biohacking communities. *Deviant Behavior*, *43*(10), 1245-1266.

Sutherland, L. A., Peter, S., and Zagata, L. (2015). Conceptualizing multi-regime interactions: The role of the agriculture sector in renewable energy transitions. *Research Policy*, *44*(8), 1543-1554.

Szarka, J. (2013). From climate advocacy to public engagement: an exploration of the roles of environmental non-governmental organizations. *Climate*, *1*(1), 12-27.

Sze, J., and London, J. K. (2008). Environmental justice at the crossroads. *Sociology Compass*, *2*(4), 1331-1354.

Szopik-Depczyńska, K., Kędzierska-Szczepaniak, A., Szczepaniak, K., Cheba, K., Gajda, W., and Ioppolo, G. (2018). Innovation in sustainable development: an investigation of the EU context using 2030 agenda indicators. *Land use policy*, *79*, 251-262.

Tan, S. Y., Taeihagh, A., and Tripathi, A. (2021). Tensions and antagonistic interactions of risks and ethics of using robotics and autonomous systems in long-term care. *Technological forecasting and social change*, *167*, 120686.

Tang, H. P. O. (2013). Recent development in analysis of persistent organic pollutants under the Stockholm Convention. *TrAC Trends in Analytical Chemistry*, *45*, 48-66.

Tankard, M. E., and Paluck, E. L. (2016). Norm perception as a vehicle for social change. *Social Issues and Policy Review, 10*(1), 181-211.

Tariq, A., Khan, A., and Bashir, A. (2020). Social work and inclusive education: Understanding the role of social workers in realizing the educational needs of children with disability. *Social Work Bulletin, 1*, 35-49.

Tarlock, A. D. (1992). The role of non-governmental organizations in the development of international environmental law. *Chi.-Kent L. Rev., 68*, 61.

Tavanti, M., and Sfeir-Younis, A. (2024). *Sustainability Beyond 2030: Trajectories and Priorities for Our Sustainable Future. Taylor and Francis.*

Taylor, B. (2001). Earth and nature-based spirituality (part I): From deep ecology to radical environmentalism. *Religion, 31*(2), 175-193.

Taylor, P. J., and Buttel, F. H. (1992). How do we know we have global environmental problems? Science and the globalization of environmental discourse. *Geoforum, 23*(3), 405-416.

Teelucksingh, C. (2019). Diverse environmentalism and inclusivity in Toronto's Green Economy. *Environmental Sociology, 5*(1), 47-58.

Teixeira, R. G. (2021). The role of international organizations in the development of international environmental law: adjusting the lenses of analysis. *Case W. Res. J. Int'l L., 53*, 237.

Telles, A. B. (2019). Community Engagement vs. Racial Equity: Can Community Engagement Work be Racially Equitable?. *Metropolitan Universities, 30*(2), 95-108.

Temper, L., Del Bene, D., and Martinez-Alier, J. (2015). Mapping the frontiers and front lines of global environmental justice: the EJAtlas. *Journal of Political Ecology, 22*(1), 255-278.

Teske, S., Morris, T., and Nagrath, K. (2020). 100% Renewable Energy for Costa Rica. *Report prepared by ISF for the World Future Council/Germany and the One Earth Foundation, USA.*

Thakur, R., and Van Langenhove, L. (2007). Enhancing global governance through regional integration. In *Regionalisation and global governance* (pp. 33-58). *Routledge.*

Thaler, T., Attems, M. S., Bonnefond, M., Clarke, D., Gatien-Tournat, A., Gralepois, M., ... and Fuchs, S. (2019). Drivers and barriers of adaptation initiatives–How societal transformation affects natural hazard management and risk mitigation in Europe. *Science of the total environment, 650*, 1073-1082.

Thiele, L. P. (1999). Environmentalism for a new millennium: the challenge of coevolution. *Oxford University Press.*

Thompson, R. G., Langemeier, L. N., Lee, C. T., Lee, E., and Thrall, R. M. (1990). The role of multiplier bounds in efficiency analysis with application to Kansas farming. *Journal of Econometrics, 46*(1-2), 93-108.

Tienhaara, K. (2010). A tale of two crises: What the global financial crisis means for the global environmental crisis. *Environmental Policy and Governance, 20*(3), 197-208.

Tol, R. S. (2014). Correction and update: The economic effects of climate change. *Journal of Economic Perspectives, 28*(2), 221-226.

Tol, R. S. J. (2009). The economic effects of climate change. *Journal of economic perspectives, 23*(2), 29-51.

Tomain, J. P. (2015). The democratization of energy. *Vand. J. Transnat'l L.*, *48*, 1125.

Tomić, T., and Schneider, D. R. (2020). Circular economy in waste management–Socio-economic effect of changes in waste management system structure. *Journal of Environmental Management*, *267*, 110564.

Tonn, B. E., and Stiefel, D. (2019). Anticipating the unanticipated-unintended consequences of scientific and technological purposive actions. *World Futures Review*, *11*(1), 19-50.

Trochim, W. M., Cabrera, D. A., Milstein, B., Gallagher, R. S., and Leischow, S. J. (2006). Practical challenges of systems thinking and modeling in public health. *American journal of public health*, *96*(3), 538-546.

Truffer, B., and Coenen, L. (2012). Environmental innovation and sustainability transitions in regional studies. *Regional studies*, *46*(1), 1-21.

Tye, M., Leong, C., Tan, F., Tan, B., and Khoo, Y. H. (2018). Social media for empowerment in social movements: the case of Malaysia's grassroots activism. *Communications of the Association for Information Systems*, *42*(1), 15.

Tzankova, Z. (2020). Public policy spillovers from private energy governance: New opportunities for the political acceleration of renewable energy transitions. *Energy Research and Social Science*, *67*, 101504.

Udawatta, R. P., Rankoth, L. M., and Jose, S. (2021). Agroforestry for biodiversity conservation. *Agroforestry and ecosystem services*, 245-274.

Ulibarri, N., Figueroa, O. P., and Grant, A. (2022). Barriers and opportunities to incorporating environmental justice in the National Environmental Policy act. *Environmental Impact Assessment Review*, *97*, 106880.

Ulloa, A. (2017). Perspectives of environmental justice from Indigenous peoples of Latin America: A relational Indigenous environmental justice. *Environmental Justice*, *10*(6), 175-180.

Upadhyaya, M. (2019). Green Web-II-Standards and Perspectives from the IUCN: Program and Policy Development in Environment Conservation Domain. *Adv Agr Environ Sci*, *2*(2), 93-95.

Upreti, B. R. (2000). *Social transformation through community forestry: Experiences and lessons from Nepal.* Retrieved October 3, 2010, academia.edu.

Vagin, S. G., Klimenko, V. A., Telegina, Z. A., and Aleksashina, T. V. (2022). Improving environmental decision-making in environmental business-management using big data and AI. *Frontiers in Environmental Science*, *10*, 951306.

Vakulchuk, R., Overland, I., and Suryadi, B. (2023). ASEAN's energy transition: How to attract more investment in renewable energy. *Energy, Ecology and Environment*, *8*(1), 1-16.

Van Breda, A. D. (2001). *Resilience theory: A literature review.* Pretoria, South Africa: South African Military Health Service, 613.

van den Bergh, J. C. (2012). Evolutionary Analysis of the Relationship between Economic Growth, Environmental Quality, and Resource Scarcity. In *Scarcity and Growth Revisited* (pp. 177-197*). Routledge.*

Van der Brugge, R., and Roosjen, R. (2015). An institutional and socio-cultural perspective on the adaptation pathways approach. *Journal of Water and Climate Change*, *6*(4), 743-758.

Van Ginkel, K. C., Botzen, W. W., Haasnoot, M., Bachner, G., Steininger, K. W., Hinkel, J., ... and Bosello, F. (2020). Climate change induced socio-economic tipping points: review and stakeholder consultation for policy-relevant research. *Environmental Research Letters, 15(2), 023001.*

Van Oss, H. G., and Padovani, A. C. (2003). Cement manufacture and the environment part II: environmental challenges and opportunities. *Journal of Industrial Ecology*, *7*(1), 93-126.

Van Rensburg, E. J. (1994). Social transformation in response to the environment crisis: The role of education and research. *Australian Journal of Environmental Education*, *10*, 1-20.

Verma, P., and Raghubanshi, A. S. (2018). Urban sustainability indicators: Challenges and opportunities. *Ecological indicators*, *93*, 282-291.

Vermaas, P., Kroes, P., van de Poel, I., Franssen, M., and Houkes, W. (2011). Ethics and Unintended Consequences of Technology. In *A Philosophy of Technology: From Technical Artefacts to Sociotechnical Systems* (pp. 103-115). Cham: Springer International Publishing.

Victor, P. A., and Jackson, T. (2012). A commentary on UNEP's green economy scenarios. *Ecological Economics*, *77*, 11-15.

Vierra, S. (2016). *Green building standards and certification systems*. National Institute of Building Sciences, Washington, DC.

Vollenbroek, F. A. (2002). Sustainable development and the challenge of innovation. *Journal of Cleaner production, 10(3), 215-223.*

von Storch, H., Tol, R. S., and Flöser, G. (2008). *Environmental crises*. Springer.

Voulvoulis, N., Giakoumis, T., Hunt, C., Kioupi, V., Petrou, N., Souliotis, I., and Vaghela, C. J. G. E. C. (2022). Systems thinking as a paradigm shift for sustainability transformation. *Global Environmental Change*, *75*, 102544.

Waisová, Š. (2015). Environmental cooperation and conflict transformation. *Central European Journal of International and Scurity Studies*, *9*(2), 92-116.

Wallace, K. J. (2007). Classification of ecosystem services: problems and solutions. *Biological Conservation*, *139*(3-4), 235-246.

Wan Rosely, W. I. H., and Voulvoulis, N. (2023). Systems thinking for the sustainability transformation of urban water systems. *Critical Reviews in Environmental Science and Technology*, *53*(11), 1127-1147.

Wang, H., Di Pietro, G., Wu, X., Lahdelma, R., Verda, V., and Haavisto, I. (2018). Renewable and sustainable energy transitions for countries with different climates and renewable energy sources potentials. *Energies*, *11*(12), 3523.

Wang, Y., Liu, J., Tang, X., Wang, Y., An, H., and Yi, H. (2023). Decarbonization pathways of China's iron and steel industry toward carbon neutrality. *Resources, Conservation and Recycling*, *194*, 106994.

Wapner, P. (1996). Globalization and the Future of Environmental Activism. *The Brown Journal of World Affairs*, *3*(2), 249-255.

Wapner, P. (2003). World Summit on Sustainable Development: toward a post-Jo'burg environmentalism. *Global Environmental Politics, 3(1), 1-10.*

Wapner, P. (2013). *Living through the end of nature: the future of American environmentalism.* MIT press.

Ward, M., Schulz, A. J., Israel, B. A., Rice, K., Martenies, S. E., and Markarian, E. (2018). A conceptual framework for evaluating health equity promotion within community-based participatory research partnerships. *Evaluation and program planning*, *70*, 25-34.

Wardani, J., Bos, J. J. A., Ramirez-Lovering, D., and Capon, A. G. (2024). Towards a practice framework for transdisciplinary collaboration in planetary health. *Global Sustainability*, *7*, e16.

Warriner, G. K., McSpurren, K., and Nabalamba, A. (2001). Social justice and environmental equity: Distributing environmental quality. *Environments*, *29*(1), 85-99.

Weichselgartner, J., and Kasperson, R. (2010). Barriers in the science-policy-practice interface: Toward a knowledge-action-system in global environmental change research. *Global environmental change*, *20*(2), 266-277.

Welford, R. (Ed.). (2013). *Hijacking environmentalism: Corporate responses to sustainable development.* Routledge.

Wheeler, H. C., and Root-Bernstein, M. (2020). Informing decision-making with Indigenous and local knowledge and science. *Journal of Applied Ecology*, *57*(9), 1634-1643.

White, A. (2010). Religious environmentalism as a vital contribution to sustainability. *Journal of Enterprising Communities: People and Places in the Global Economy*, *4*(3), 268-278.

Whiting, K., Carmona, L. G., and Sousa, T. (2017). A review of the use of exergy to evaluate the sustainability of fossil fuels and non-fuel mineral depletion. *Renewable and Sustainable Energy Reviews*, *76*, 202-211.

Whittaker, R. J., and Fernández-Palacios, J. M. (2007). Island biogeography: ecology, evolution, and conservation. Oxford University Press.

Whittington, C. (2023). Collaboration and partnership in context. In *Social Work (pp. 240-246).* Routledge.

Wiedmann, T., Lenzen, M., Turner, K., and Barrett, J. (2007). Examining the global environmental impact of regional consumption activities—Part 2: Review of input–output models for the assessment of environmental impacts embodied in trade. *Ecological economics*, *61*(1), 15-26.

Wilkinson, J., Goff, M., Rusoja, E., Hanson, C., and Swanson, R. C. (2018). The application of systems thinking concepts, methods, and tools to global health practices: an analysis of case studies. *Journal of evaluation in clinical practice*, *24*(3), 607-618.

Willems, J. C. (1997). On interconnections, control, and feedback. *IEEE Transactions on Automatic control*, *42*(3), 326-339.

Williams, E., Kahhat, R., Allenby, B., Kavazanjian, E., Kim, J., and Xu, M. (2008). Environmental, social, and economic implications of global reuse and recycling of personal computers. *Environmental science and technology*, *42*(17), 6446-6454.

Williams, F. (2018). Intersectionality, gender and social policy. In *Handbook on gender and social policy* (pp. 37-54). Edward Elgar Publishing.

Williams, J. H., Jones, R. A., Haley, B., Kwok, G., Hargreaves, J., Farbes, J., and Torn, M. S. (2021). Carbon-neutral pathways for the United States. *AGU advances*, *2*(1), e2020AV000284.

Wolbring, G., and Burke, B. (2013). Reflecting on education for sustainable development through two lenses: Ability studies and disability studies. *Sustainability*, *5*(6), 2327-2342.

Wright, C., Kiparoglou, V., Williams, M., and Hilton, J. (2012). A framework for resilience thinking. *Procedia Computer Science*, *8*, 45-52.

Xu, X., Huang, S., An, H., Vigne, S., and Lucey, B. (2021). The influence pathways of financial development on environmental quality: new evidence from smooth transition regression models: renewable *and Sustainable Energy Reviews*, *151*, 111576.

Xu, Y., Ge, W., Liu, G., Su, X., Zhu, J., Yang, C., ... and Ran, Q. (2023). The impact of local government competition and green technology innovation on economic low-carbon transition: new insights from China. *Environmental Science and Pollution Research*, *30*(9), 23714-23735.

Yaga, D., Mell, P., Roby, N., and Scarfone, K. (2019). Blockchain technology overview. *arXiv preprint arXiv:1906.11078.*

Yang, D., Liu, D., Huang, A., Lin, J., and Xu, L. (2021). Using a LEAP scenario modeling, critical transformation pathways and socio-environmental benefits of energy substitution. *Renewable and Sustainable Energy Reviews*, *135*, 110116.

Yearley, S. (2013). Social movements and environmental change. In *Social theory and the global environment* (pp. 150-168). Routledge.

Yegina, N. A., Zemskova, E. S., Anikina, N. V., and Gorin, V. A. (2020). Model of consumer behavior during the digital transformation of the economy. *Industrial Engineering and Management Systems*, *19*(3), 576-588.

Yen, N. T. H., and Dung, N. P. (2023). Marine pollution management in response to plastic waste under international law and national law–challenges and implications for developing countries: a case study of Vietnam. *Asia Pacific Journal of Environmental Law*, *26*(2), 162-193.

Yigitcanlar, T. (2015). Smart cities: An effective urban development and management model?. *Australian Planner*, *52*(1), 27-34.

York, R., and Bell, S. E. (2019). Energy transitions or additions?: Why a transition from fossil fuels requires more than the growth of renewable energy. *Energy Research and Social Science*, *51*, 40-43.

Young, O. R. (2003). Environmental governance: the role of institutions in causing and confronting environmental problems. *International Environmental Agreements*, *3*, 377-393.

Young, O. R. (2008). The architecture of global environmental governance: bringing science to bear on policy. *Global Environmental Politics*, *8*(1), 14-32.

Zaccaï, E. (Ed.). (2007). *Sustainable consumption, ecology and fair trade* (Vol. 156). London: Routledge.

Zhang, R., and Hanaoka, T. (2022). Cross-cutting scenarios and strategies for designing decarbonization pathways in the transport sector toward carbon neutrality. *Nature communications*, *13*(1), 3629.

Zhang, X., Li, Y., Shi, K., and Feng, Y. (2022). How do environmental technology standards affect the green transformation? New evidence from China. *International Journal of Environmental Research and Public Health*, *19*(10), 5883.

Zhang, X., Shu, K., Rajkumar, S. A., and Sivakumar, V. (2021). Research on deep integration of application of artificial intelligence in environmental monitoring system and real economy. *Environmental Impact Assessment Review, 86*, 106499.

Zhao, S., Wu, C., Hong, H., and Zhang, L. (2009). Linking the concept of ecological footprint and valuation of ecosystem services–a case study of economic growth and natural carrying capacity. *International Journal of Sustainable Development and World Ecology, 16*(2), 137-142.

Zhou, Y., Zheng, S., Lei, J., and Zi, Y. (2023). A cross-scale modeling and decarbonization quantification approach for navigating Carbon Neutrality Pathways in China. *Energy Conversion and Management, 297*, 117733.

Zhu, T., and Wang, L. (2020). *State Energy Transition*. Springer Singapore.

Zimmerman, M. E. (2005). Integral ecology: A perspectival, developmental, and coordinating approach to environmental problems. *World futures, 61*(1-2), 50-62.

Zinchenko, V. V., Boichenko, M. I., Polishchuk, O. S., and Drach, I. I. (2021). Transformations of global cooperation processes as a strategy of sustainable social development in the context of research and education. In *E3S Web of Conferences* (Vol. 277, p. 02001). EDP Sciences.

Zuhdi, U. (2014, April). Using multipliers analysis to get another perspective related to the role of ICT sectors in the national economy of Indonesia: 1990-2005. In *Journal of Physics: Conference Series* (Vol. 495, No. 1, p. 012051). IOP Publishing.

Zürn, M. (1998). The rise of international environmental politics: a review of current research. *World Politics, 50*(4), 617-649.

Index

A

B

C

D

E

F

G

H

I

L

M

N

O

P

R

S

T

U

W